Computerized Maintenance Management Systems Made Easy

Computerized Maintenance Management Systems Made Easy

How to Evaluate, Select, and Manage CMMS

Kishan Bagadia

McGraw-Hill

New York Chicago San Francisco Lisbon London Madrid
Mexico City Milan New Delhi San Juan Seoul
Singapore Sydney Toronto

The McGraw·Hill Companies

CIP Data is on file with the Library of Congress

1 2 3 4 5 6 7 8 9 0 DOC/DOC 0 1 2 1 0 9 8 7 6

ISBN 0-07-146985-0

The sponsoring editor for this book was Kenneth P. McCombs and the production supervisor was Richard C. Ruzycka. It was set in Century Schoolbook by International Typesetting and Composition. The art director for the cover was Anthony Landi.

Printed and bound by RR Donnelley.

This book is printed on recycled, acid-free paper containing a minimum of 50% recycled, de-inked fiber.

Contents

Preface

This book is intended to be a comprehensive practical reference on computerized maintenance for plant engineering personnel and others concerned with plant operations and maintenance management. It applies to industrial plants, utilities, hospitals, and building maintenance. In fact, it applies to all operations involved with equipment and inventory management. Because of its practical usefulness, this book will serve as an excellent handout in maintenance management courses, seminars and workshops.

As global competition intensifies, operational executives are adopting new concepts and technologies to increase productivity and quality. Yet, many overlook the most important operational support system—the maintenance organization. While maintenance is traditionally underutilized, proactive organizations are increasingly recognizing the impact of maintenance on increased equipment availability and utilization, on improved quality, and thus on resultant competitiveness and profitability. Maintenance is becoming a focus for competitive advantage.

This book covers all aspects of CMMS starting with overview of CMMS, leading to justifying, specifying, evaluating, selecting and implementing a CMMS. It also covers auditing and optimizing your existing CMMS making the book useful to readers who may or may not have a CMMS. This book is not about some ideas and principles; you will be able to develop specific actions to be implemented as soon as you finish reading the book. The book will help readers improve maintenance productivity, machine quality, reduce downtime and increase overall profits.

Kishan Bagadia

Acknowledgments

This book is dedicated to my late mother who would have been very happy to witness this achievement.

I am thankful to my father who has inspired me to succeed all my life.

I would like to thank my wife, Uma, who has supported me while I spent time writing this book.

My thanks are also due to all three of our sons who have been a tremendous help in making this project a success. The oldest, Nikhil, has been instrumental in organizing this book. The second son, Nishant, helped with proofreading and correcting the manuscript. And our youngest, Avi, helped with some typing in spite of his homework and other teenage priorities.

My special thanks to all the editors and other staff members at McGraw-Hill for doing a great job.

Finally, very special thanks to the other authors who have contributed valuable information to complete this book.

Computerized Maintenance Management Systems Made Easy

Chapter

1

Maintenance Management

Maintenance can be defined as the orderly control of activities required to keep a facility in an as-built condition, with the ability to maintain its original productive capacity. Maintenance management simply involves managing the control of maintenance activities.

A good maintenance management system makes equipment and facilities available. Availability means the production team can demand and receive any item such as light, power, air, gas, heating, cooling, or machine tools when it is needed. If the required equipment or service is down, or if the machine stops short of completing a job, time and money are wasted. A good maintenance management system helps accomplish minimal downtime.

The fundamental steps of maintenance management are simple and beneficial to pursue. The basic steps of maintenance management are request, approval, plan, schedule, performing work, recording data, accounting for costs, developing management information, updating equipment history, and providing management control reports. The following is a brief review of the events that occur during each of these steps. Also see Fig. 1.1.

- *Request.* Requests to perform maintenance work may be transmitted in different ways—verbal, written, electronic, and the like.
- *Approval.* Maintenance supervisors often handle simple jobs (small expenditures); however, large expenditures require approvals from several levels of management.
- *Plan.* Ensure resources are available (material, labor, tools, equipment, and so forth). The planning step can take many forms. For

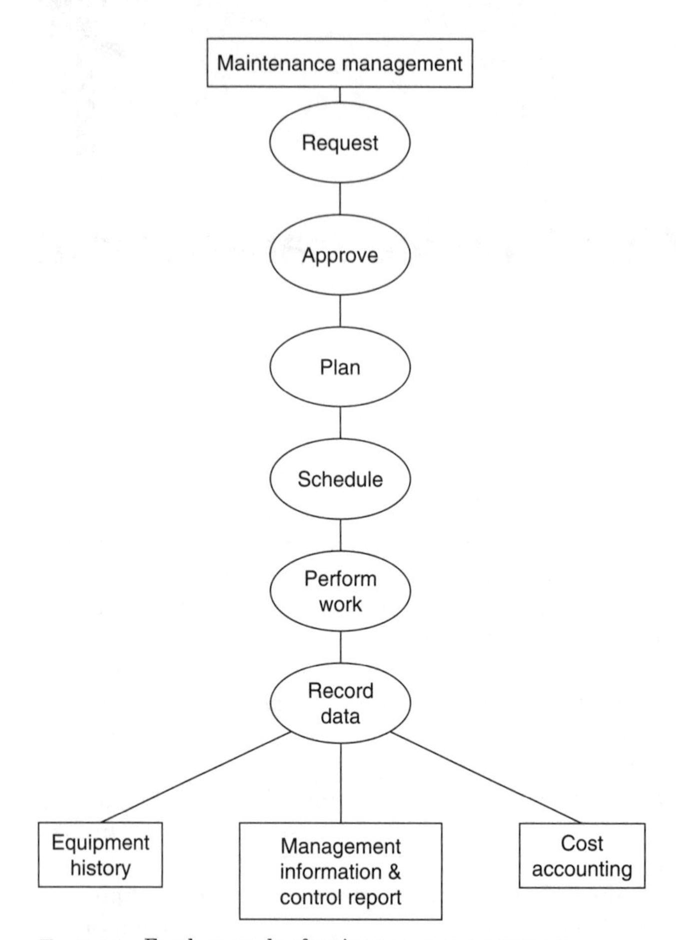

Figure 1.1 Fundamentals of maintenance management.

example, the maintenance supervisor can give verbal orders to craftspeople, and if a planner is used, he/she will prepare a work order.

- *Schedule.* Scheduling involves three factors:
  - *Priorities.* Priority codes are based on established criteria or on the importance of the equipment and the type of work.
  - *Job assignment.* Jobs are assigned to maintenance technicians.
  - *Follow-up.* Follow up to make sure the job is being done.
- *Performing work.* The work is performed based on information received by craftspeople. Sometimes they receive little or no instructions and sometimes they are provided with detailed procedures.
- *Record data.* The data recording may vary from simple listing of the actual hours to keeping comprehensive records of material charges,

equipment identification, work assigned and performed, and other pertinent data.

- *Accounting for costs.* It is important to know where your money is used, and what you purchase with the money.
- *Developing management information.* This may involve providing facts on current work including costs, accumulated data, equipment identification, productivity, budgets, and scheduling.
- *Updating equipment history.* History records might include little or no data, or might show updates of all equipment, records, equipment use, downtime, maintenance labor, and material costs.
- *Management control reports.* As management information is developed, control reports covering expenditures, performance, backlog, and equipment data, for example, are generated regularly to summarize the results of the maintenance function. These reports are very important for plant managers to make appropriate and intelligent decisions.

Chapter

2

Overview of CMMS

Computerized Maintenance Management System (CMMS)

A *computerized maintenance management system* (CMMS) is a computer software program designed to assist in the planning, management, and administrative functions required for effective maintenance. These functions include the generating, planning, and reporting of *work orders* (WOs); the development of a traceable history; and the recording of parts transactions.

CMMS is not just a means of controlling maintenance. It is now used as a means to ensure the high quality of both equipment condition and output.

CMMS offers core maintenance functionalities. It is not limited to manufacturing; it is applicable to facilities, utilities, fleet, and other types of organizations where equipment/asset are subject to wear, and repairs are done to them. A CMMS usually includes equipment management, *preventive maintenance* (PM), labor tracking, WO, planning, scheduling, inventory control, and purchasing. CMMS usually do not include financial or human resource management (other than basic cost recording and personnel information). You can, however, integrate them with financial and human resource applications.

Enterprise asset management (EAM) evolved from CMMS. With EAM, the CMMS functionality is extended to include financial modules such as accounts payable, advanced cost recording, and advanced human resource management.

CMMS Modules

A basic CMMS includes:

A. Equipment data management

B. Preventive maintenance

C. Labor

D. Work order system

E. Scheduling/planning

F. Vendor

G. Inventory control

H. Purchasing

I. Budgeting

These or other modules may work independently or may be integrated. For example, a CMMS that links the equipment data and WO modules can automatically insert equipment information into a WO as soon as you enter the equipment ID. The result is a quicker, more accurate WO containing consistent data.

The need and use of a CMMS is not specific to any one industry or type of application. Any industry requiring equipment and/or asset maintenance is a potential candidate for using a CMMS. A CMMS is becoming more attractive as more maintenance personnel have become computer literate and price of hardware and software have dropped significantly. Companies are also investing in CMMS as they are designed to support the requirements of ISO 9000, other regulatory agencies, and are a key part of the *total productive maintenance* (TPM) and other modern maintenance philosophies.

The following pages display a number of screens to give you better understanding of how a CMMS should appear and function. This document is written without reference to any particular CMMS and may be used as an impartial guide to assist you in determining your CMMS requirements.

Every item on every screen will not be discussed. Some of the key elements and functionalities will, however, be discussed.*

Equipment management

This module is used to define all pieces of equipment/asset, spare parts, run time, and safety procedures, and to set up schedules for PM. You can add new pieces of equipment to the database, define relationships among equipment, enter and track equipment locations.

---

*Screenshots taken from Webworks by TERO Consulting.

You can maintain complete maintenance history on equipment through WOs, which enables you to make decisions on replacing or repairing equipment. This kind of reporting can save time in analysis during an equipment life cycle.

Equipment module interacts with PM, labor, WO, inventory, and purchasing. See Fig 2.1.

Equipment number. Each piece of equipment is assigned a unique ID (or equipment number).

Procedures. Used to define PM, safety, or any other procedures. Each procedure is just typed once and assigned a code. These codes eliminate the need to repeatedly type the information in other sections, that is, in situations where the same procedure is being used by a number of equipment.

Hierarchy. You can create a parent-child relationship by specifying which equipment it belongs to (its parent), and/or which equipment belongs to it (its children).

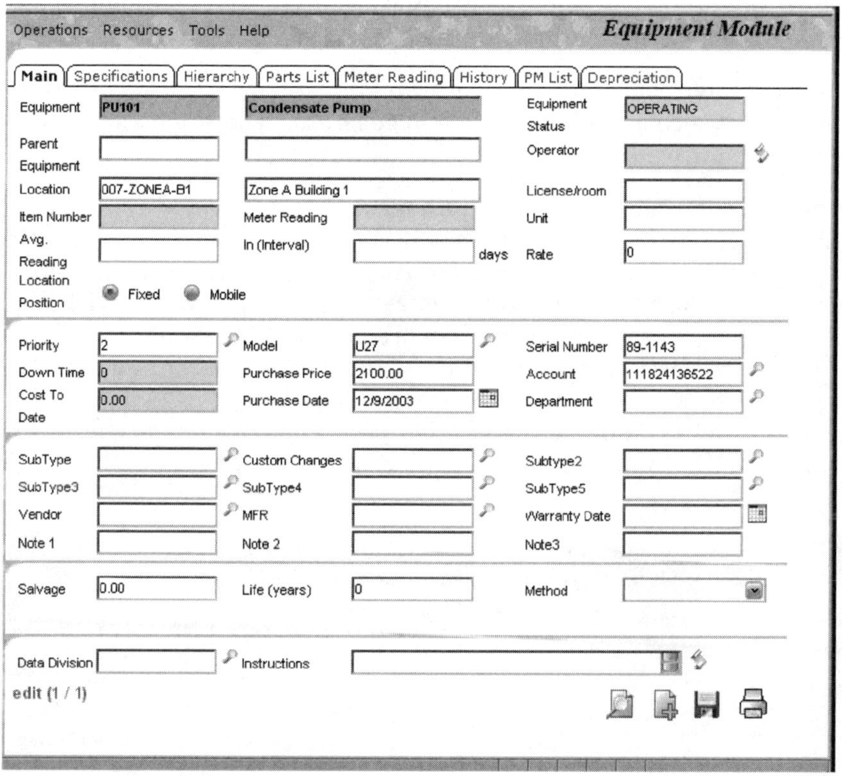

Figure 2.1 Equipment data screen.

Priority. This indicates the criticality of this equipment. This is an important field because it is used to determine which equipment to fix first in case of a problem. For example, each equipment is assigned a number from 1 to 10 (1 being least critical), so in case of an emergency, an equipment that has a No.10 priority, will be scheduled for repair before others. (For scheduling purpose, job priority is also taken into consideration.)

Run Time. This stores information regarding run time in units of miles, hours, and so on. This is useful in cases where you schedule the PM functions based on the run time rather than on the calendar time. For example, change oil every 3000 mi.

The run time reading requires periodic updates. The frequency at which you update the reading depends on the level of activity. This could vary from real time to a few times a day to once a week. The PM scheduling is done based on this reading. Also, the reading can be updated manually, semiautomatically, or automatically. Full automation is obtained by interfacing CMMS directly to run time measurement system, for example, a building control system. See Fig. 2.2.

Spare parts. This contains information on all spare parts recommended to be stored for this equipment. See Fig. 2.3.

Auxiliary equipment. Specify, store, and track information on auxiliary equipment such as motor, pump, and generator. There should be provision for multiple units.

Cost. Every time work is done on a piece of equipment, the cost (material, labor, and outside) is computed via WO module and recorded in equipment module. It usually keeps track of maintenance dollars spent on that piece of equipment *year to date* (YTD) and *life to date* (LTD). This information is vital for performing a replacement analysis.

It is very typical in maintenance environment where the maintenance manager feels certain machines are not worth repairing and should be replaced. Most CMMS do not have *machine replacement analysis* (MRA) built in the software. However, CMMS has the maintenance cost data

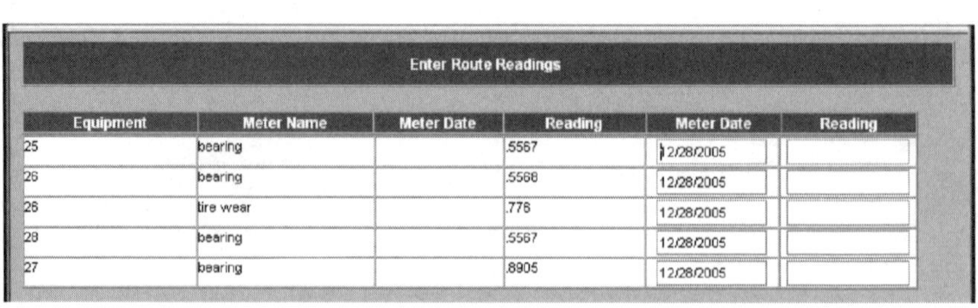

Enter Route Readings

Equipment	Meter Name	Meter Date	Reading	Meter Date	Reading
25	bearing		.5567	12/28/2005	
26	bearing		.5568	12/28/2005	
26	tire wear		.776	12/28/2005	
28	bearing		.5567	12/28/2005	
27	bearing		.8905	12/28/2005	

Figure 2.2 Equipment run time information.

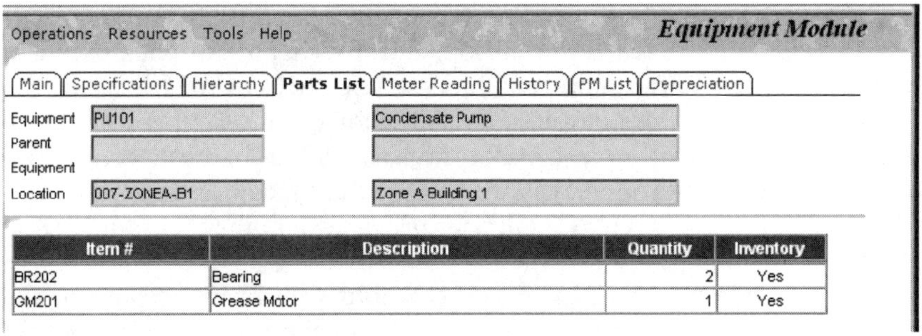

Operations Resources Tools Help — *Equipment Module*

Main | Specifications | Hierarchy | **Parts List** | Meter Reading | History | PM List | Depreciation

Equipment: PU101 — Condensate Pump
Parent Equipment:
Location: 007-ZONEA-B1 — Zone A Building 1

Item #	Description	Quantity	Inventory
BR202	Bearing	2	Yes
GM201	Grease Motor	1	Yes

Figure 2.3 Spare parts information.

that is needed to run the MRA. Some CMMS have depreciation and salvage analysis as shown in Fig. 2.4.

For details on how to run a machine replacement analysis, see App. 2A, "Machine Replacement Analysis."

Links. You can attach *computer-aided drawing* (CAD), schematics, pneumatics, and scanned documents to equipment records. This is a very effective way to store, view, and print popular image formats including digital photos.

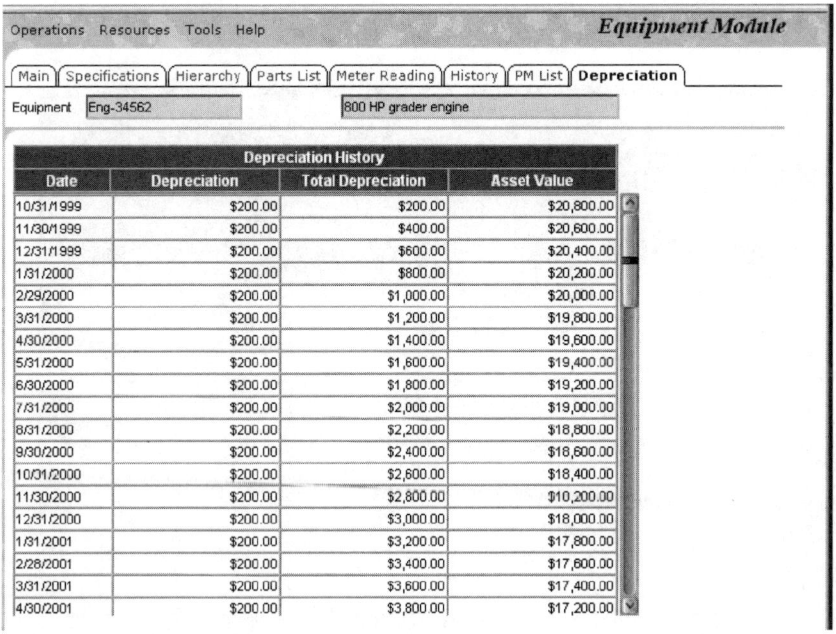

Operations Resources Tools Help — *Equipment Module*

Main | Specifications | Hierarchy | Parts List | Meter Reading | History | PM List | **Depreciation**

Equipment: Eng-34562 — 800 HP grader engine

Depreciation History			
Date	Depreciation	Total Depreciation	Asset Value
10/31/1999	$200.00	$200.00	$20,800.00
11/30/1999	$200.00	$400.00	$20,600.00
12/31/1999	$200.00	$600.00	$20,400.00
1/31/2000	$200.00	$800.00	$20,200.00
2/29/2000	$200.00	$1,000.00	$20,000.00
3/31/2000	$200.00	$1,200.00	$19,800.00
4/30/2000	$200.00	$1,400.00	$19,600.00
5/31/2000	$200.00	$1,600.00	$19,400.00
6/30/2000	$200.00	$1,800.00	$19,200.00
7/31/2000	$200.00	$2,000.00	$19,000.00
8/31/2000	$200.00	$2,200.00	$18,800.00
9/30/2000	$200.00	$2,400.00	$18,600.00
10/31/2000	$200.00	$2,600.00	$18,400.00
11/30/2000	$200.00	$2,800.00	$10,200.00
12/31/2000	$200.00	$3,000.00	$18,000.00
1/31/2001	$200.00	$3,200.00	$17,800.00
2/28/2001	$200.00	$3,400.00	$17,600.00
3/31/2001	$200.00	$3,600.00	$17,400.00
4/30/2001	$200.00	$3,800.00	$17,200.00

Figure 2.4 Depreciation and salvage analysis.

These images can be viewed and/or printed with WOs and *purchase orders* (POs). With scanning option, hand-written sketches or documents from instruction manuals can become part of CMMS.

By attaching CAD, equipment/asset information, and construction drawings, specifications are constantly available for both emergency and day-to-day situations. An awareness of the exact location and condition of specific items in a facility improves the ability to manage any situation.

If it is a web-based CMMS, you can add web links as part of the record. For example, you can link equipment manufacturer's Web site to access line diagrams, schematics, operations and maintenance (O/M) information pertinent to that piece of equipment. See Fig. 2.5.

Safety and compliance. A maintenance management function of an organization can contribute to safety and compliance with regulatory requirements. There are regulatory agencies such as:

Occupational Safety and Health Administration (OSHA)

Environmental Protection Agency (EPA)

Food and Drug Administration (FDA)

International Standard organization (ISO)

Joint Commission on Accreditation of Healthcare Organization (JCAHO)

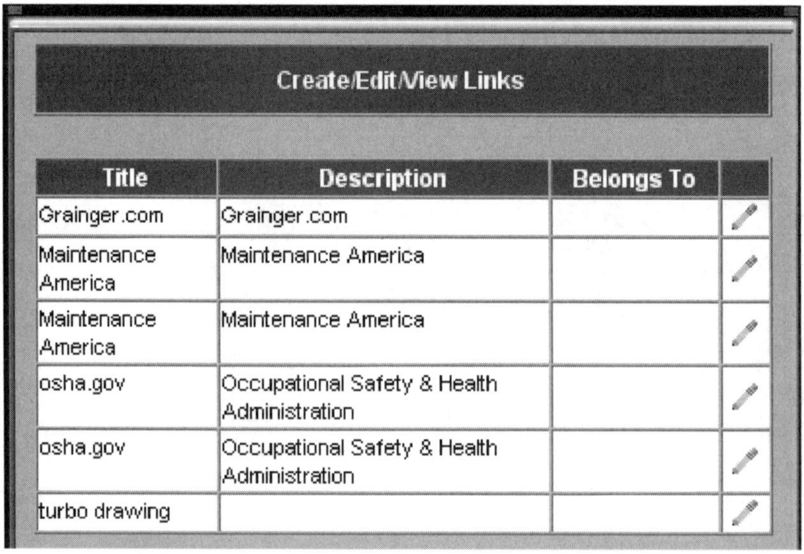

Create/Edit/View Links

Title	Description	Belongs To	
Grainger.com	Grainger.com		
Maintenance America	Maintenance America		
Maintenance America	Maintenance America		
osha.gov	Occupational Safety & Health Administration		
osha.gov	Occupational Safety & Health Administration		
turbo drawing			

Figure 2.5 Attachments and web links.

These are most commonly encountered agencies; however, there are other agencies with compliance regulations in certain types of industries. Each has its own regulations and compliance requirements. By understanding the requirements in the context of maintenance, compliance with these regulations will be easier to achieve and maintain.

A well-designed, properly implemented CMMS will help ensure regulatory compliance. However, organizations that neglect this will find themselves in regulatory violations and pay heavy penalties.

You should study the agency requirements that apply to you. Typical requirements include, proper record keeping, routine maintenance and inspection, lock out/tag out procedures and so forth. CMMS should print lock out/tag out instructions on every applicable WO to ensure that maintenance technicians are provided with appropriate instructions.

Some regulations require an audit trail and date/time stamp of activities, for example, routine PM. A CMMS can provide this with the help of mobile technology.

Compliance versus noncompliance. Compliance is not difficult. It is simply implementing a CMMS with basic functions such as equipment data, PM, WO, and inventory. With proper training to your employees you can easily achieve and maintain regulatory compliance. It is a small investment that really pays off.

Failure to comply can be very expensive. Companies have been charged millions of dollars for noncompliance.

To see how a CMMS can help with regulatory compliance issues, see App. 2B, "ISO/QS Compliance Case Study."

Warranty. Most CMMS have a field for warranty data entry. However, it does not do anything more than storing that data. Every time you issue a WO for a piece of equipment, the CMMS should flag a warning if that equipment is still under warranty. Then you should be easily able to access the vendor information that provides the warranty. This feature can save significant amount of money as often maintenance technicians end up fixing things that are still under warranty due to lack of information.

Condition monitoring. This function allows you to set an unlimited number of measurement points for each piece of equipment. These measurement points make it possible to determine if the equipment is operating within safe parameters. Predictive maintenance is a good example. CMMS should be capable of generating WOs if measurements are outside acceptable limits.

Equipment reports. Most CMMS packages provide a number of standard reports. This module generates reports such as:

- Cost (equipment number, date installed, original cost, LTD cost)
- Equipment failure (equipment number, description, failure codes: with report queries you can get details of failures)
- Equipment hierarchy (equipment number, location, description. Top-level equipment and its children are listed)
- Failure count (displays number of failures for each piece of equipment in a given time period)
- Equipment warranty (equipment number, description, warranty expiration date)
- Equipment availability (equipment number, description, times available during a specified period)
- Special tools (equipment number, description, special tools required)
- Meter reading (equipment number, description, date, current meter reading, previous date, and meter reading)
- Location (equipment number, description, location, department)
- Summary (equipment number, description)
- Master (all fields)

Preventive maintenance (PM)

Introduction. Most equipment need periodic maintenance to ensure uninterrupted efficient operations. You can use the PM module to create PM records and generate WOs. PM records contain task description, material, and labor information. A PM record specifies work to be performed regularly based on calendar time or run time such as hours or miles.

The PM interacts with equipment, WO, labor, and inventory modules. See Fig. 2.6.

Procedures. Entering PM procedures could be very time consuming. You can predefine PM procedures (PM task details, labor, parts, and tools requirements). You can then use this procedure for any equipment you desire. This saves a tremendous amount of data entry time, particularly if a number of equipment use the same procedure.

These procedures are not limited to PM tasks. They can be safety procedures or any other set of instructions.

Priority. Priority assigned to this PM function. This priority should be automatically transferred to the WO generated for this PM function. That becomes the job priority.

Frequency. The system stores information on how often PM is scheduled, either by calendar time (weekly, monthly, and so on) or by run time

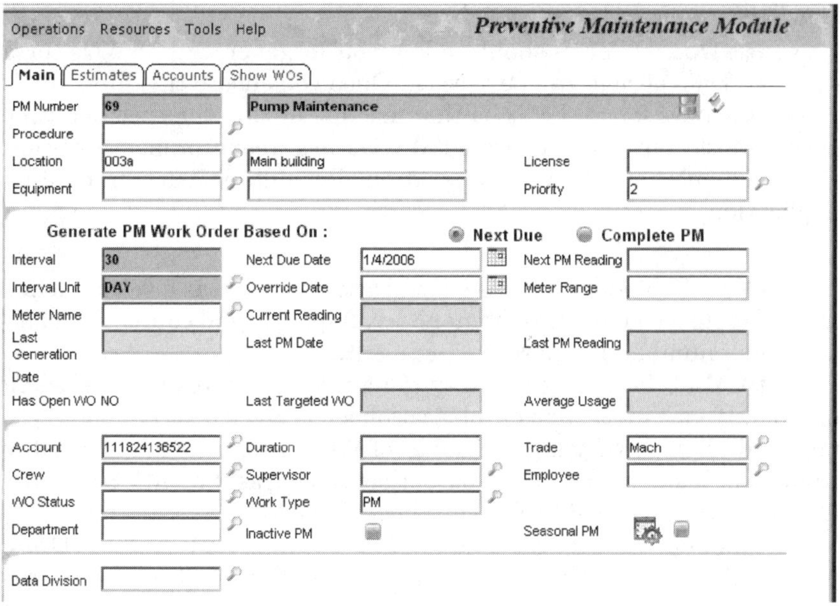

Figure 2.6 Preventive maintenance.

(miles, hours, and so on). Based on the frequency, the system schedules PM jobs.

Seasonal PM. CMMS should accommodate seasonal PM frequency adjustments. For example, you do not have to do any maintenance on a lawn mower in winter and a snow blower in summer.

Parts. Enter parts required to perform this PM job including quantity of each part. This list should be automatically transferred to the WO when one is generated for this task.

Labor. Enter craft categories assigned to this PM job (e.g., mechanic and electrician) and their corresponding estimated hours to complete this job. This list should be automatically transferred to the WO for this PM function. When the WO is generated, individual employees within that craft category will be assigned to the job. Some CMMS have the capability to assign employees to PM task as well.

Outside contractor. Sometimes PM jobs are contracted to outside vendors. CMMS should have the ability to keep track of all details including their schedules. You want to maintain complete history of PM jobs done regardless of who performed it.

Links. This feature works the same way as specified in equipment module.

Functions

Generate WOs. CMMS should allow you to generate PM WOs for a specified time period, say next week. Once you choose this function, it generates WOs for all PM jobs that are due during the specified time period. Most companies generate PM WOs on a weekly or monthly basis. The frequency of PM WO generation depends on each individual application.

Route PM. Route-based PM is a list of activities such as lubrication and inspection to be done by maintenance craftspeople and equipment operators on a number of different equipment.

Most CMMS lack the capability to handle the routes in an efficient manner. If each inspection or lubrication job is given a WO number, you may end up with very large number of WOs. From the stand point of equipment history, it is a desirable option, as long as the CMMS provides for an efficient method to close all those WOs.

The other option is to create one WO for the whole route. The WO will print a complete list. However, when you close the WO, history may not be maintained on each piece of equipment. You have to evaluate your application and choose the appropriate option.

PM reports

- PM list (list of all PM jobs that are due in a specified period of time)
- PM labor (details including money spent on PM labor)
- PM material (details including money spent on PM material)
- PM labor projection (forecast labor requirements for a specified period. This report can be further broken down by craft category or individuals)
- PM material projection (forecast material requirements for a specified period.)

Labor

This module keeps track of maintenance employees' information including the hourly wage rates for cost calculations upon closing or updating a WO. You can create craft categories representing group of employees, for example, mechanics. You can also enter shift, vacation, and sick time information, which can be used for scheduling purpose. This module can be used to generate time cards for maintenance employees. See Fig. 2.7.

Labor module interfaces with PM, WO, planning, and scheduling.

Operations Resources Tools Help *Labour Module*

Main | Additional Info | Equipment History

Employee ID	TR1007	Last Name	Smith		
Title		First Name	Joe		
Is Employee	Yes / No				
Address	3545 Cherrywood Lane				
City	Milwaukee	State	WI	Country	USA
Zip	53222	Phone	414-787-9845	Fax	
URL					
Email	jsmith@yahoo.com			Email Notification	
Craft	Mach	Hourly Rate	16	Hire Date	12/2/2002
Location		Department		lab1	
Employee #		lab3		lab4	
Division					

Figure 2.7 Employee information.

Craft code. This represents the type of craft category like plumber, mechanic, carpenter, and so forth. For example,

ELEC	Electrician
MECH	Mechanic
CARP	Carpenter

Hourly rate. This is the employee's hourly wage rate. Software should accommodate *overtime* (OT), double time, call in, or other applicable rates.

Links. You can attach employees' photographs to their record.

Labor reports

- Labor overtime (employee ID, name, OT taken, OT refused)
- Labor vacation (employee ID, name, hours earned, hours scheduled, hours used)
- Labor skill (employee ID, name, craft code, skill level)
- Labor summary by account number (employee ID, WO number, hours queried by account number)
- Labor summary by employee ID (employee ID, WO number, hours queried by employee)
- Labor summary by calendar (employee ID, WO number, hours queried by specific time period)

- Labor summary by equipment (employee ID, WO number, hours queried by specific equipment)
- Labor productivity (can be used to monitor performance based on estimated and actual hours)
- Labor master (all fields)

Work order system (WO). WO is heart of a CMMS. This module would allow you to generate, print, and complete WOs. It stores all preventive and corrective maintenance WOs while work is either going on, or planned for in the future. It can also serve as a powerful tool for cost estimating. Once you enter the labor, material, and outside cost information, the system will calculate the estimated cost of the project.

You should be able to accommodate both unplanned work (emergency) and planned work (scheduled). WOs can be created for a piece of equipment/asset. The cost can be charged to a particular account or multiple accounts.

All the work requests are generated through work request module or WO module. If a work request module is used, each request gets converted into a WO. The information stays in there until the work is finished. Upon completion, the record is transferred into the work history file.

"Mobile technology with CMMS" section of this book shows how a CMMS with mobile technology can minimize the steps in WO completion process. You can have a completely paperless WO system as shown in Fig. 2.8.

WO request module. Organizations where many different people submit work requests present a problem of getting work requests submitted efficiently. Sometimes people drop off hand-written notes. Sometimes they call

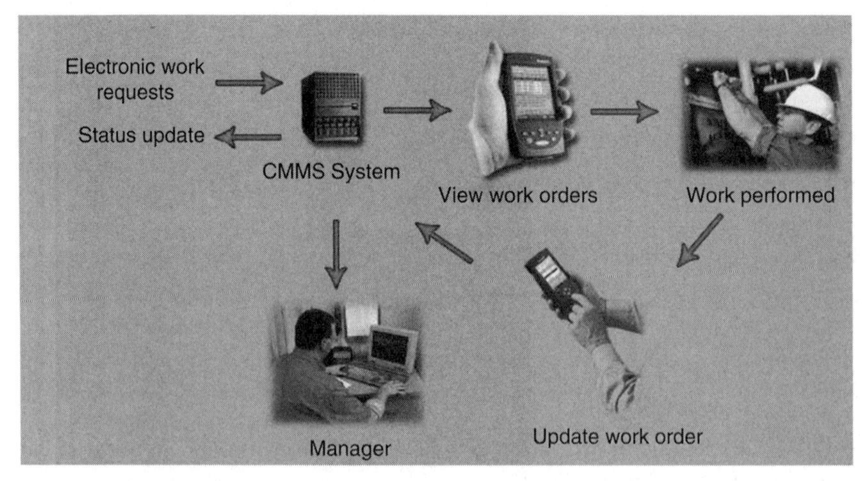

Figure 2.8 Paperless work order system.

and other times a staff person will see a maintenance technician repairing equipment and verbally request another job that would only take a minute.

Whatever the method, this kind of work request causes problems. Often there is no record of the work requested, requestor, or the time of the request. Notes get lost. Conversations may be remembered differently. Phone calls can occur during busy times when the information is not recorded accurately, and phone messages left on voice mail can be unclear or incomplete. As a result of all this, work does not get done leading to frustration.

By moving to electronic work requests, both the requester and the maintenance department have a clear communication. See Fig. 2.9.

With this kind of set up, both requestors and maintenance staff are satisfied. Requestors are not running down the hall or trying to reach maintenance personnel over the phone. Maintenance technicians get the complete information they need. When the status of the work request is posted electronically, requestors can check on the progress at any time; maintenance technicians spend more time completing tasks instead of tracking work requests.

WO interacts with equipment, PM, labor, inventory, and purchasing modules.

WO module. See Fig. 2.10.

Priority. Priority code for this job. Assign the importance of the job (you may use a number from 1 to 10 or any other scheme your CMMS provides). Most CMMS provide a priority field but do not use it for any purpose other than storing this information. Priority should be used for scheduling aid.

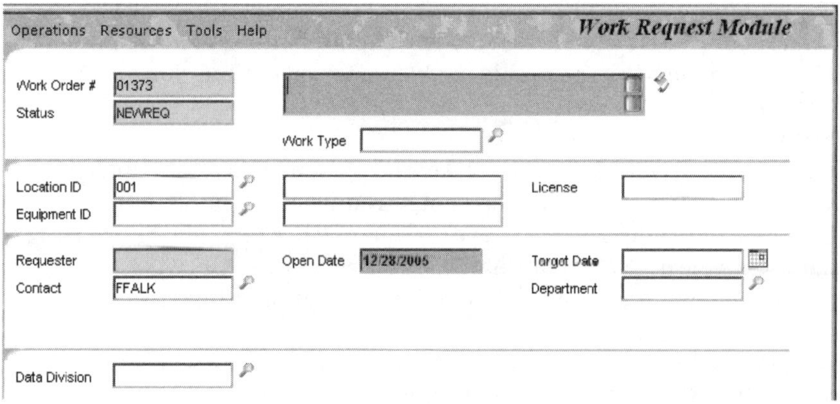

Figure 2.9 Work order request screen.

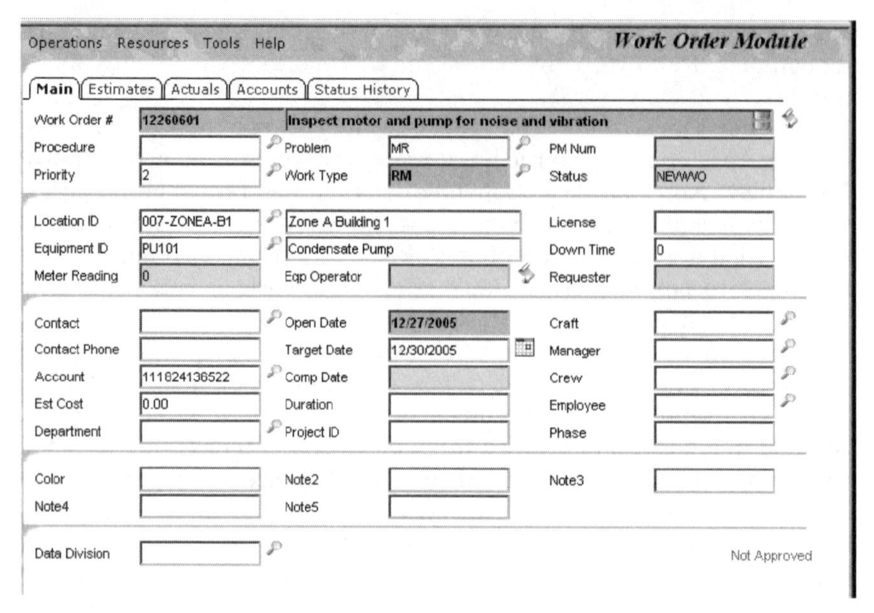

Figure 2.10 Work order screen.

Status. Indicates the status of this WO. For example, waiting for approval and waiting for material. This makes it easier for any one to access the system and view the status of a job.

Category. Work category for this job, for example, PM, emergency, repair, and project. These should be user defined.

Failure code. Failure code suitable for this job. Explains what is wrong with the equipment/asset. These should be user defined.

Action code. Action code for this job, explains the action taken, that is, what was done to fix the problem. Over a period of time, you build useful history.

Labor. Allows you to enter the ID of the person, who is scheduled to perform this job, and the estimated hours required to complete the job. You can enter multiple crafts. For example, a machine installation job might require an electrician, a mechanic, and a carpenter. Once the job is completed, you can record actual labor time spent. See Fig. 2.11.

Material. Allows you to enter material required for each WO. Upon completion or as material is withdrawn, you can record that in the system. See Fig. 2.12.

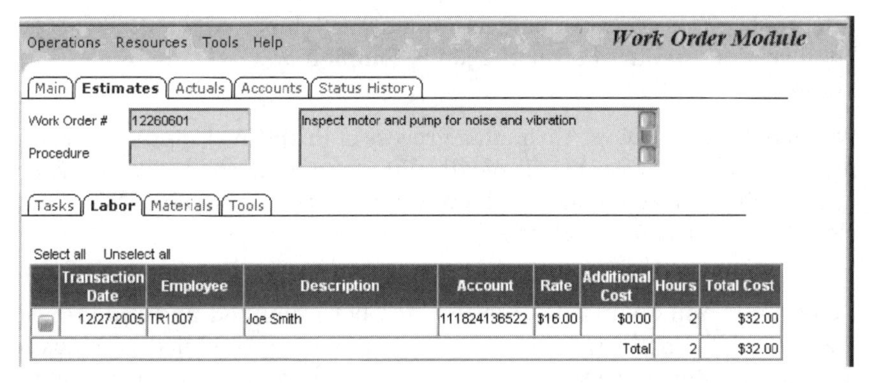

	Transaction Date	Employee	Description	Account	Rate	Additional Cost	Hours	Total Cost
	12/27/2005	TR1007	Joe Smith	111824136522	$16.00	$0.00	2	$32.00
						Total	2	$32.00

Figure 2.11 Work order labor hours.

Material cost. Based on the material used for this WO and the unit cost information entered in the inventory file, the program computes the total cost of material.

Labor cost. Based on the actual time spent by each technician and the corresponding hourly rate entered in the labor file, the program computes the total cost of labor.

Outside cost. Any outside cost incurred for this job. This provides the ability to track maintenance costs performed by outside contractors. This cost is added to the material and labor cost when calculating the total cost.

Total estimated cost. This is sum of estimated material, labor, and outside costs.

Operations Resources Tools Help
Work Order Module
Main Estimates Actuals Accounts Status History Complete
Work Order # 00003
Procedure LOF
Lube ,oil, filter
Tasks Labor Materials Tools
Select all Unselect all

	Issue Date	Item Number	Description	Account	Quantity	Issue Price	Extension
	3/23/2000	M-AF-2324	Air filter Chev 1/2 ton 77-99		1	$17.89	$17.89
	3/23/2000	M-FL-2345	Oil filter Chev 1/2 ton 76-97		1	$11.67	$11.67
					Total		$29.56

Figure 2.12 Work order material.

Total cost. This is the sum of actual material, labor, and outside costs. Total cost is computed after the job is completed.

WO completion. You go through a process of initiating, approving, completing, and closing a WO. Typically, WOs can be closed individually or as a batch. Batch completion allows you to select a number of WOs and close them with one touch of key.

Downtime. You should be able to track both planned and unplanned downtime. Planned downtime means equipment is scheduled to be available for maintenance work. Unplanned downtime means the equipment goes down unexpectedly.

Reducing unplanned downtime saves you money. You can track downtime, analyze downtime trends, and take action to reduce unplanned downtime in the future. You can analyze the data to find the total downtime for a piece of equipment, cause of downtime, and the cost of downtime to your company.

This kind of analysis will help you make decisions on rescheduling PMs and replacing existing equipment.

WO reports

- WO parts shortage (WO number, part number quantity required, quantity on hand, quantity on order, quantity short)
- Active WO (all WOs pending during a specified time period)
- Overdue WO (all WOs that are overdue at a specified date)
- WO material requirement (part number, WO numbers, quantity required, quantity on hand, quantity on order)
- WO labor requirements (craft categories, WO #, required and available hours)
- WO detail (WO number, date, work description)
- Downtime summary (WO number, lost operation hours, cost)
- Account history (account number, WO numbers, category, failure code, cost)
- Activity (category, number of WOs, percentage of WOs by categories)
- Performance [WO number, date required, date completed, difference (days), percent completed in time]
- Cost summary (WO number, labor cost, material cost, outside cost, total cost)
- Labor summary (WO number, hours, labor cost)

- Material summary (WO number, part number, material cost)
- Equipment history (equipment number, WO numbers, category, failure code, cost)
- Cost variance (WO number, estimated cost, actual cost, percent variance)

Certain reports are easier to read in graphics format compared to tabular form. Figures 2.13 and 2.14 display examples of graphics reports.

Planning/scheduling

The greatest potential for improvements in maintenance productivity, quality, costs, and responsiveness is to determine:

- What work is required to satisfy a request?
- How can the work be performed in the most efficient manner?
- What resources are required to accomplish the required work?
- When should the work be scheduled for execution based on work priority and the availability of resources?

Someone plans all maintenance work in some way. Unfortunately when it is not formally planned at the appropriate level, results are unfavorable. Breakdown maintenance modes consistently produce:

- Low labor productivity
- Excessive work delays

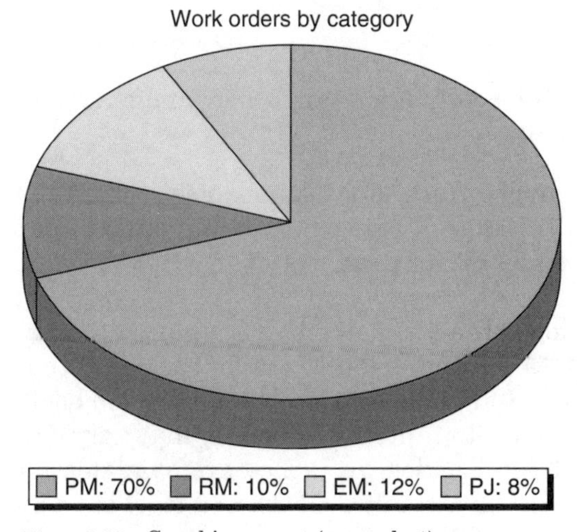

Figure 2.13 Graphics report (example 1).

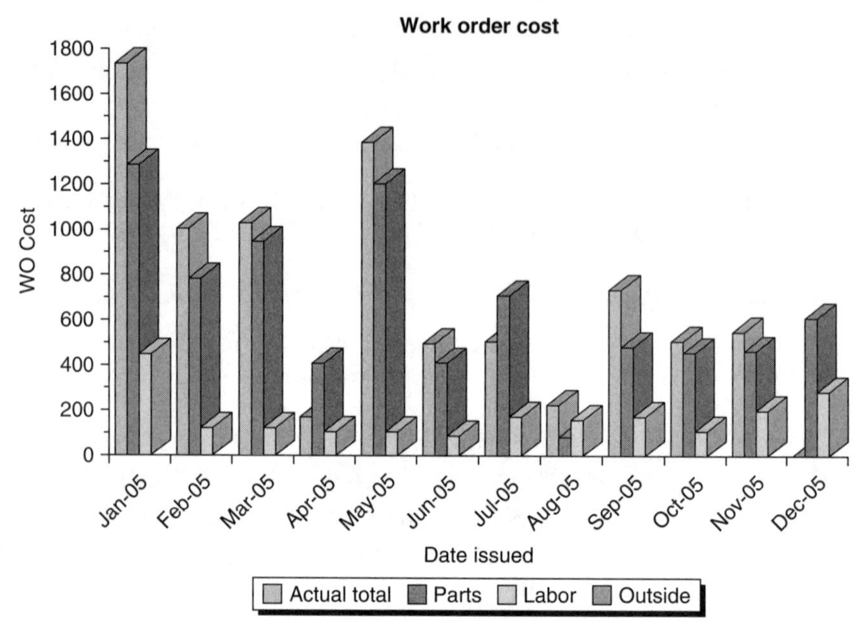

Figure 2.14 Graphics report (example 2).

- Overtime work
- Continuous crises
- Unmanageable backlog
- Inventory high in value but low in turnover and service levels
- Employee frustration
- Customer dissatisfaction
- Production downtime which leads to uncertain order promising and delivery

In contrast, when maintenance work is predetermined, accurately prioritized, and aligned with required resources, maintenance work control and customer service are within specification.

Maintenance backlog. Backlog is the estimated hours required to complete the approved WOs. Jobs are estimated in hours for each craft that is to perform work on the WO. Total hours for each craft become backlog hours for that particular craft. Definitions of backlog may vary, but usually WOs are not considered backlog until they are available for scheduling. Each operation will have a different desired level of backlog. Most plants use a minimum of 80 hs of work per craftsman.

Maintenance organization can use backlog as an important tool to identify the following:

- Workload distribution by area, priority, and so forth
- Balance of personnel to workload
- Determine if work should be contracted outside
- Determine which work should be deferred
- Need for resource adjustment within the maintenance organization

Priority system and backlog. Priorities affect backlog. Priorities are essential for objectivity in assigning order of execution. Priority of a WO can be determined by multiplying the equipment criticality (Fig. 2.15) by the job priority (Fig. 2.16). This is called *relative importance factor* (RIF). Jobs will be scheduled from higher to lower RIF. For example:

$$\text{RIF} = \text{job priority} \times \text{equipment criticality}$$

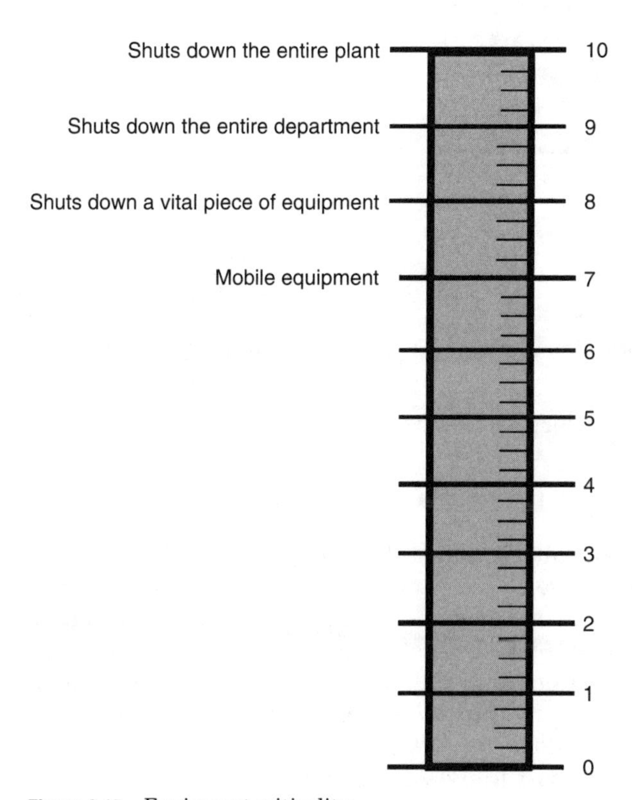

Figure 2.15 Equipment criticality.

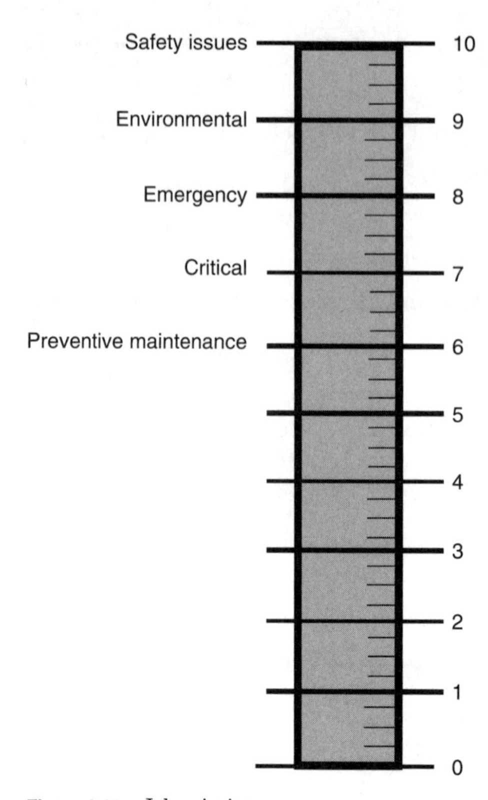

Figure 2.16 Job priority.

Example

- Say, WO 1 has a job priority of 10 with an equipment criticality of 7: RIF = 70
- And WO 2 has a job priority of 9 with an equipment criticality of 10: RIF = 90
- WO 2 has a higher priority than WO 1

CMMS allows you to plan and schedule WOs based on RIF, available resources such as manpower, parts, and equipment. Planning/scheduling module typically does not have an input screen. The information required for this purpose comes from other modules such as equipment (availability), PM (schedule, material, and labor requirements), labor (manpower availability), inventory (material availability), and purchasing (parts on order information)

A typical scheduling report is shown in Fig. 2.17.

WO Number	CAT	PRI	CR	RIF	Date required	Equip	Name	Description
100204002	EM	10	10	100	10/2/2004	HV0021	HVAC Unit 1	Check for motor
100304005	EM	10	9	90	10/3/2004	HV0025	HVAC Unit 2	Doesn't function at all
100204009	R	8	8	64	10/4/2004	LT-005	Lathe engine	Blew a gasket*
100304008	R	7	8	56	10/10/2004	AC2045	Air compressor	Check bearing/bushing
100204012	PM	6	9	54	10/5/2004	FT004	Ford truck	Change oil
100404015	PM	6	9	54	10/6/2004	CR-448	Crane	Teardown & inspection
100104005	PM	6	8	48	10/8/2004	HT562	Hoist	General inspection
100404018	PM	6	8	48	10/6/2004	CR-449	Crane	Lubrication
100404019	PM	6	8	48	10/6/2004	CR-450	Crane	Lubrication
100204021	PM	6	7	42	10/5/2004	DR054	Radial drill	Check oil level

Total mechanic hours required:	90
Total mechanic hours available:	80

* Material not available

Figure 2.17 Work order planning/scheduling.

The WOs are sorted by RIF (higher to lower). It identifies WOs that have material shortage allowing you to make a decision either to reschedule those WOs or acquire material from alternate sources. The report also compares available versus required manpower (mechanic in this example). Available manpower is short of required. You can either reschedule some of the jobs or assign overtime.

Vendor

This module stores information about your suppliers. These are companies you purchase parts, equipment, or services (such as maintenance or outside contractors) from. It contains information such as name, address, phone, fax, email, and payment terms for each vendor. It also allows you to print mailing labels for the vendors. See Fig. 2.18.

Vendor reports

- Address labels (print address labels for all or selected vendors)
- Contact reports (vendor ID, name, contact, address)
- Cost variance (percentage of cost increases)

Inventory control

Managing inventory is an important part of maintaining any plant/facility. Inventory control keeps track of items in stock, indicates when

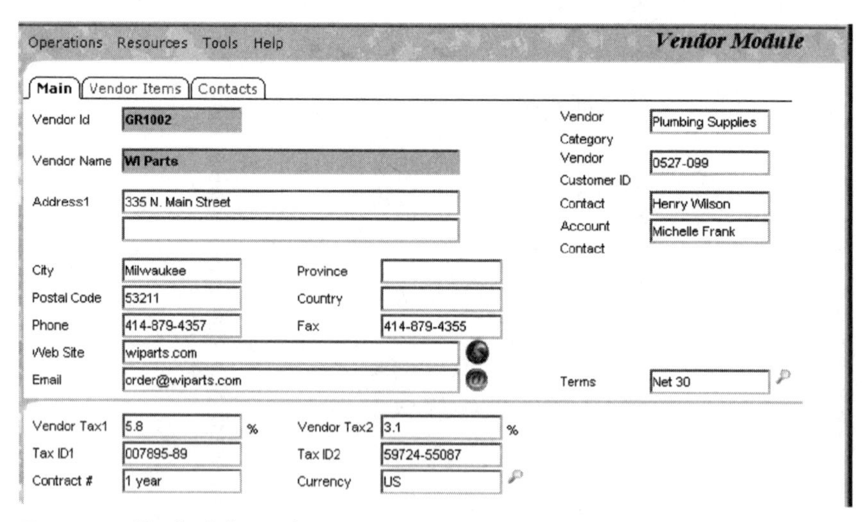

Figure 2.18 Vendor information.

stock falls to user-defined reorder points, creates POs to restock needed items. It keeps track of locations for each part.

This module keeps track of quantities in stock. You can specify a *reorder point* (ROP) and an *economic order quantity* (EOQ) for each item. When the parts reach ROP, CMMS generates a requisition for those parts and a PO upon approval.

Most CMMS treat ROP and EOQ as input. The calculations of these can be complex and there are many different models for these computations.

See App. 2C, "Spare Parts Inventory Management," for some discussion on this topic.

Functions include

Issues. Keep track as items are depleted from inventory. Items are typically withdrawn against WOs.

Receipts. Add items to inventory as parts are received. Items are typically received against purchase orders (POs).

Allocations (reserve). Items are reserved for WOs.

Inventory module interacts with equipment, WO, planning/scheduling, and purchasing. See Fig. 2.19.

Description. A brief description of the part, for example, pump, motor, and belt. Usually there is a provision to enter additional details elsewhere in the program.

Substitute (alternate). An alternate part that can be substituted for this item if necessary.

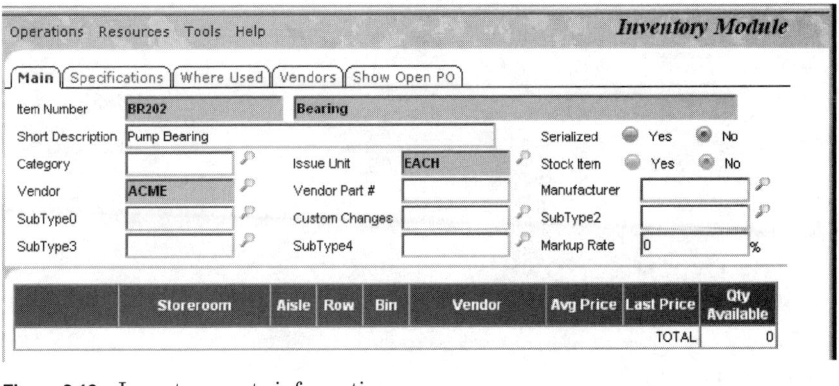

Figure 2.19 Inventory parts information.

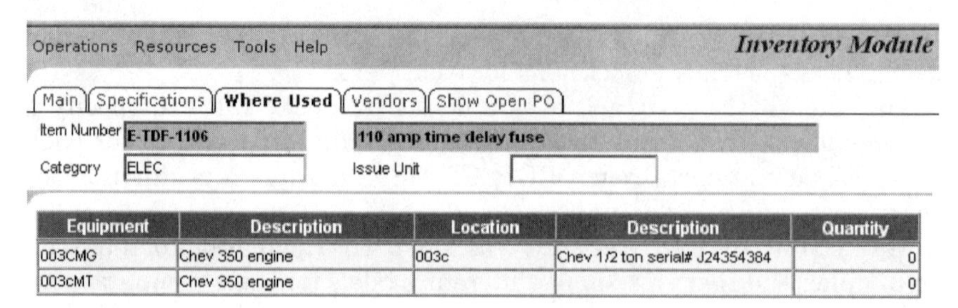
Operations Resources Tools Help *Inventory Module*

Main | Specifications | **Where Used** | Vendors | Show Open PO

Item Number: E-TDF-1106 — 110 amp time delay fuse

Category: ELEC — Issue Unit:

Equipment	Description	Location	Description	Quantity
003CMG	Chev 350 engine	003c	Chev 1/2 ton serial# J24354384	0
003cMT	Chev 350 engine			0

Figure 2.20 Where used parts.

Where used. Enter all equipment numbers where this part is used. This is very useful information in case there is a machine breakdown and the part is not in stock. You can potentially use the part from a different piece of equipment (if that equipment is less critical) and replenish when part is received at a later date. See Fig. 2.20.

Location. Identifies part location. Allows tracking of inventory stocked at multiple locations. With web-based CMMS, keeping track of inventory at multiple plants/facilities is much easier. See Fig. 2.21.

Unit of measure. The unit of measurement (e.g., each, gallon, feet, and rolls).

Reorder level. Quantity that represents the ROP for this part number. When the quantity on hand falls at or below the reorder level, you should order more parts.

Storeroom

Item Number: BR202 — Description: Bearing

StoreRoom:

Aisle: — Row

Vendor: ACME — Vendor P

Account:

Select a Storeroom

Storeroom:

Description:

Storeroom	Description

Stock Level: 0 — Last Pric

Reserved Quantity: 0 — Average

Qty Available: 0 — Quoted Price: 56.00 — Issue Unit: EACH

Quantity On Order: 0

Reorder Method: Min / Max — EOQ Reorder Point

Max Stock / EOQ: — Min Stock / Reorder Point:

Figure 2.21 Parts location.

Quantity allocated. The quantity reserved for various jobs.

Transfers. Easy transfer functionality should be provided between locations if you have multiple stockrooms.

Vendor. Stores useful information for each vendor associated with this part, for example, lead-time, manufacturer part number, unit price, and reorder quantity. One of the vendors is usually identified as the primary vendor. POs are issued automatically to the primary vendor with manual override capability.

Cross-reference. Inventory should have a provision to store manufacturers' and vendors' part numbers. A cross-reference function should be available between company (your), manufacturer, and vendor part numbers.

Open WOs. Should show all open WOs needing a particular part.

Open POs. Should show all open POs for a particular part.

Physical count. CMMS should have a provision for physical count. It can be a manual (paper-based) system or one using mobile technology with CMMS.

Links. Works the same way as specified in equipment module.

Inventory reports

- Below ROP (part number, description, quantity on hand, ROP, quantity on order)
- Location (part number, description, location, shelf/bin)
- Inventory on hand (part number, description, quantity, unit cost, extended cost)
- Part cost (part number, description, vendor, cost)
- Obsolescence (part number, description, date last used, date last received)
- Summary (part number, description)
- Parts usage history (part number, description, number of transactions, quantity used, quantity on hand, extended cost)
- Parts usage history by equipment (same as the preceding point, queried by equipment number)
- Inventory transactions (part number, addition, subtraction, quantity)

- Parts allocation (part number, description, WO number, quantity available, quantity allocated)
- Print labels (part number, description, location)

Purchasing

The purchasing module enables you to create and process purchase requisitions, and POs. You can order, receive, and track both materials and services.

There are two steps to purchasing with CMMS:

- Requisition
- Actual issuing of the POs

First, a requisition is generated for all parts below ROP (When quantity on hand plus quantity on order is less than the reorder point, the system sends a flag requesting that the item be reordered.).

This becomes a request for purchasing items. Someone must then approve this request. Once approved, the PO is issued. With one key command, a CMMS should be able to issue all POs that are due and approved.

You should be able to revise a PO before the vendor processes it.

Some CMMS have provision to receive price quotes. A request for quotes is sent out to vendors in the database. This is how a CMMS saves you money in purchasing parts. The vendors are aware of competition and you get the best prices.

Purchasing module interacts with inventory, vendors, and WOs. See Fig. 2.22.

Dates. Should have the ability to enter and track date required, date promised, and date received.

PO status. User-defined status codes.

Receiving. When parts are received against a PO, quantity and costs are updated. Inventory activity is recorded as well. You can receive stock and nonstock items. You can also receive and track POs for services rendered by contractors, for example, repair on a fork truck. System should allow partial receipts.

Blanket POs. The CMMS should have the ability to issue blanket POs.

Currency. With recent advances in globalization, it is becoming more important to have currency management within CMMS. It should provide the ability to convert currencies of various countries.

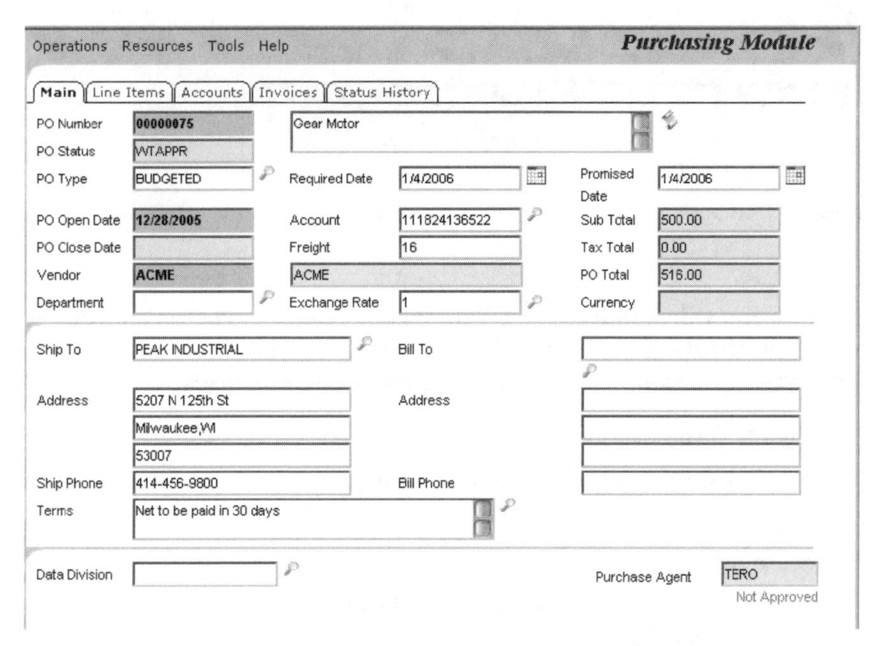

Figure 2.22 Purchase order information.

Closing POs. Like WOs, you go through a process of initiating, approving, receiving, and closing a PO. Typically, POs can be closed individually or as a batch.

Links. Works the same way as described in equipment module.

PO reports

- Open PO (PO number, PO date, vendor, required date, promised date, cost)
- Parts received (PO number, PO date, part number, quantity ordered, quantity received)
- Cost variance (PO number, part number, price promised (or last price), price charged, percentage of variance)
- Vendor performance (PO number, date promised, date delivered, late days, percentage variance)

Budgeting

This module allows you to set up various budget accounts and allocate funds to each account (e.g., repairs, PM, projects). Every time a

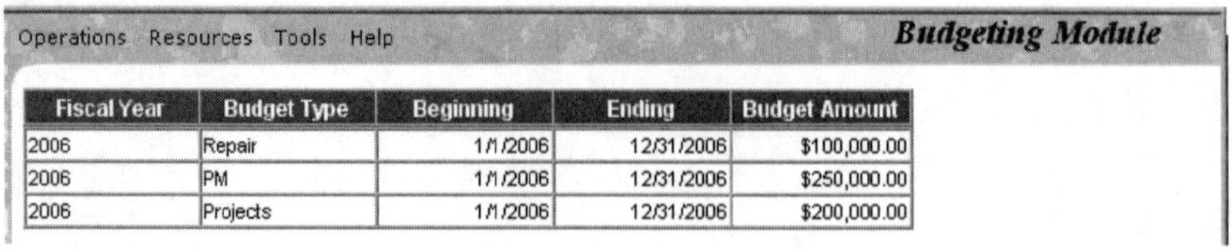

Operations Resources Tools Help *Budgeting Module*

Fiscal Year	Budget Type	Beginning	Ending	Budget Amount
2006	Repair	1/1/2006	12/31/2006	$100,000.00
2006	PM	1/1/2006	12/31/2006	$250,000.00
2006	Projects	1/1/2006	12/31/2006	$200,000.00

Figure 2.23 Budget accounts.

transaction takes place (e.g., WO, parts issue), the cost is charged to the appropriate account number. CMMS keeps track of money spent toward each account (Fig. 2.23). It serves two purposes:

1. To control the expenditure
2. To help set up budget for the following year

Budget reports

- Budget cost (account number, cost allocated)
- Labor variance (account number, cost allocated, actual cost, percentage of variance)
- Material variance (account number, cost allocated, actual cost, percentage of variance)
- Budget variance (account number, cost allocated, actual cost, percentage variance)

Additional features

Security. Security is a very important function of a CMMS package.

A multilevel password protected security system can define exactly what each user is permitted to access, edit, or run. A good security system allows the following:

- Make different screens for different users (e.g., a person just making a work request would view a very simple WO screen, as opposed to the maintenance supervisor who will view the whole WO screen).
- Make fields visible or hidden (you do not want everyone to view employees' pay rates).
- Make fields editable or noneditable (this provides database security).
- Make fields mandatory.
- Validate data entry against predefined criteria.

Duplicating a record. With CMMS application, there is a frequent need for duplicating a record, editing, and then saving as a new record to save retyping effort. For example, let us say, you have 15 pumps that are identical. Instead of typing equipment information 15 times, you can create one record, duplicate it, change serial number and location, and then save it as a new equipment record. This applies to similar PM and safety procedures and inventory parts as well.

User customizable reports. This option allows you to customize reports without any programming knowledge. You are not dependent on the vendor for report modifications, in turn saving you time and money. With this option you can:

- Modify reports and forms
- Create new reports just the way you want them
- Interface with other programs such as accounting and purchasing, even if they are running on a different operating system
- Create American standard code for information interchange (ASCII) file formats for universal data import/export

User customizable screens. This option allows you to customize screens without any programming knowledge. With this option you can:

- Modify a screen
- Change name (legend) of a field
- Change the size of a field
- Change the position of a field
- Add or delete fields
- Modify context sensitive help (This is very important since you are allowed to modify fields. If you modify a field and cannot change corresponding help, it could create major problems.)

Emerging technologies

Web-based CMMS. PC-based CMMS started in early 1980s. Initially, they were DOS based and available on a single PC. They gradually progressed to *local area network* (LAN)- and *wide area network* (WAN)-based systems commonly referred to as *client server technology* (CST). Recent explosion of web-based technology has put CMMS application in a different horizon.

The technology. A LAN is a group of computers and associated devices that share a common communications line or wireless link and typically share the resources of a single processor or server within a small geographic area (e.g., within an office building, a manufacturing plant, or facility). Usually, the server has applications and data storage that are shared in common by multiple computer users. A LAN may serve as few as two or three users (e.g., in a home network) or as many as thousands of users.

Most LANs are confined to a single building or group of buildings. However, one LAN can be connected to other LANs over any distance via telephone lines and radio waves. A system of LANs connected in this way is called a WAN.

With a web-based technology, you can access your data securely from anywhere in the world. All you need is internet access. In fact, you do not install anything. You just log in through your internet browser and a full-featured application comes up for your use.

Some of the problems with CST-based CMMS are that it takes forever to install, crashes your servers, is expensive and time consuming to upgrade, and requires expensive hardware. It poses the challenge of multiple databases (multiple plants and facilities) and synchronizing data between locations, not to mention day-to-day maintenance and support. All these hidden costs add up in a hurry. See Figs. 2.24, 2.25, and 2.26.

Recently, the web-based CMMS model has taken off because it solves these problems.

Web-based CMMS is available in two different ways. Acquire the software and install on your server or lease it. The leasing option is commonly

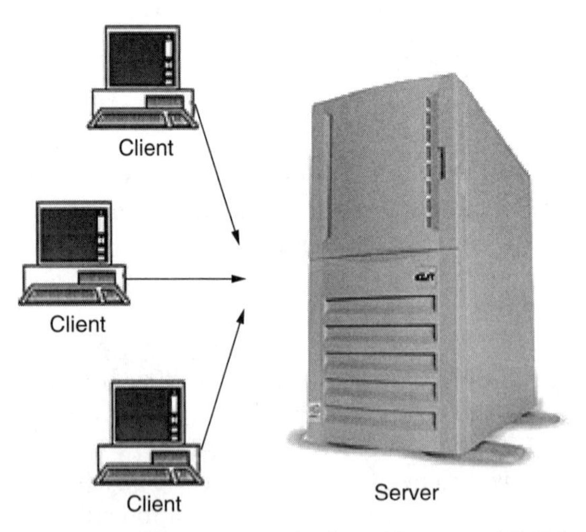

Figure 2.24 Client server technology. (*Courtesy of TERO Consulting.*)

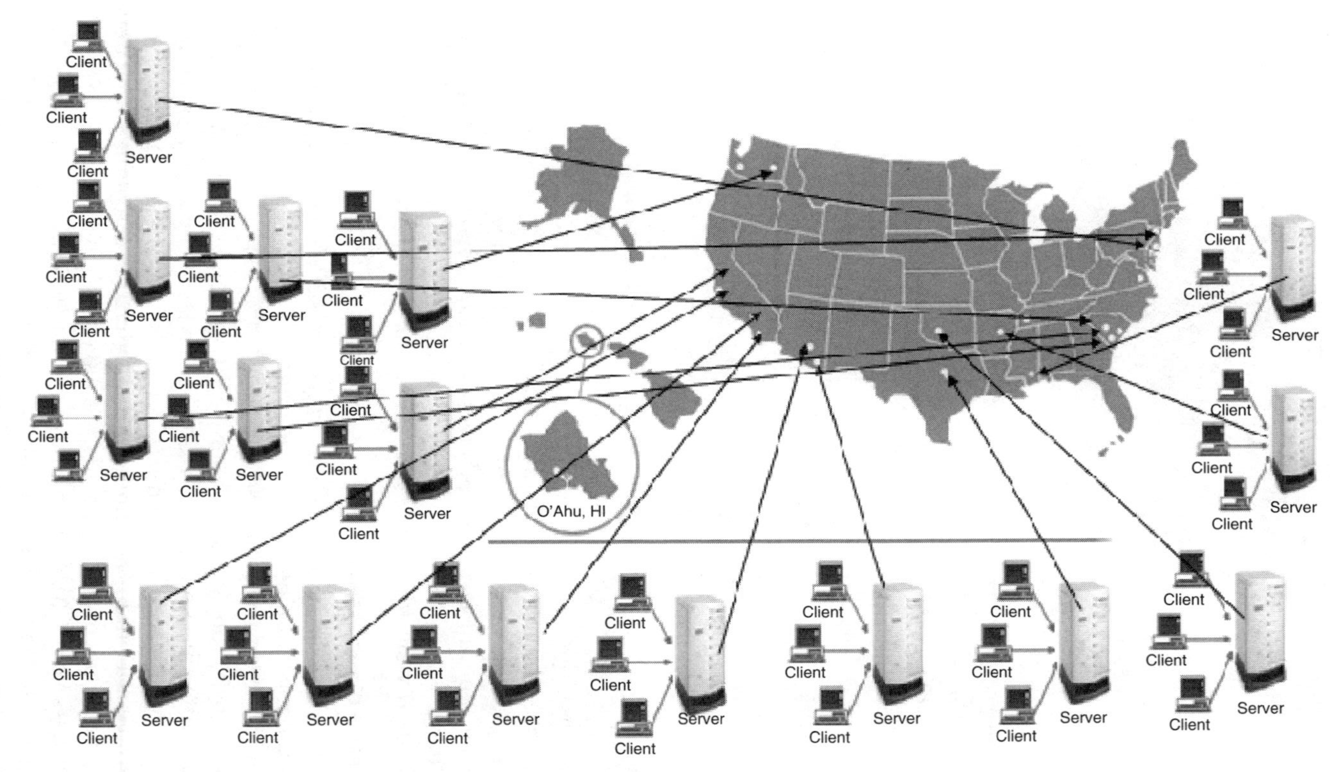

Figure 2.25 National client server setup. (*Courtesy of TERO Consulting.*)

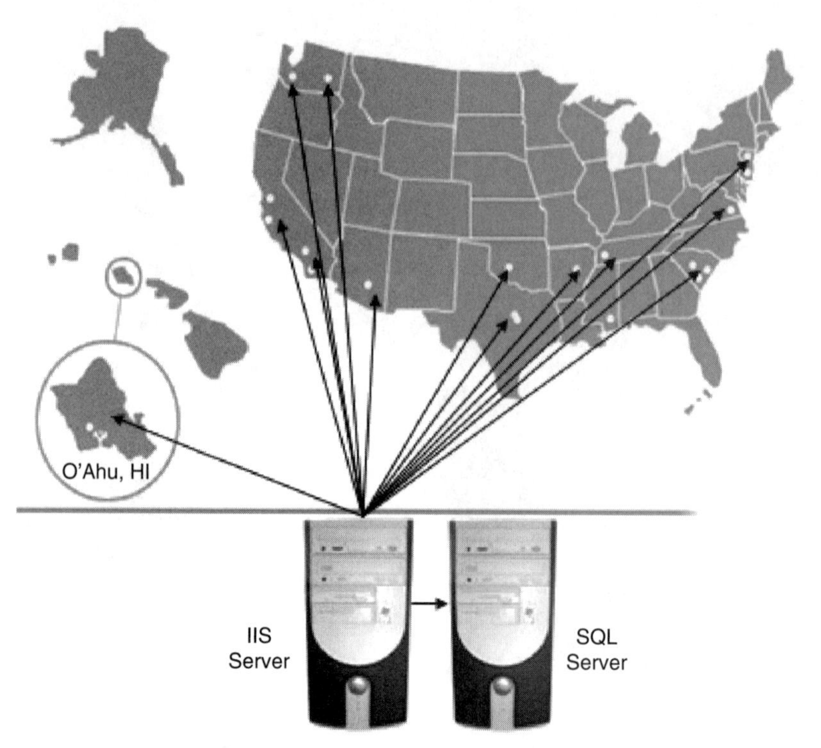

Figure 2.26 Web-based architecture. (*Courtesy of TERO Consulting.*)

known as *application service provider* (ASP) option. With ASP option, rather than buying the software, you pay a rental fees for it. The CMMS is installed on vendor's server, it is available to you through any web browser, and all support and maintenance is handled by the service provider.

Let us review why organizations are moving to web-based CMMS.

Benefits of web-based CMMS

It can be used anywhere and anytime. Web-based CMMS allows maintenance managers and supervisors to use the program wherever there is an Internet connection. It allows managers to order parts, review work requests, and generate WOs from anywhere. You can work from California, London, or Australia. As long as you have an Internet connection, a user ID, and a password, the CMMS is available to you.

If you get a call in the middle of the night from the plant with a breakdown or repair problem, you do not have to go to work right away. You can access your CMMS first and see if you can solve the problem without going to work—for example, reviewing history to see if the same problem was solved earlier and directing your people accordingly.

Imagine being stranded out of town. Your crew would be waiting for their work assignment. Check in a hotel with internet access and schedule your WOs.

Lower overall cost. Another advantage of web-based CMMS software is its lower overall cost.

Licensing fee. Cost is traditionally higher with CST.

Installation cost. With CST, you have to install CMMS on both client PCs and server, which is lot more labor intensive. Also, typically this installation has to be done by *information technology* (IT) people who may not always be available as they have other projects and their own priorities. So, there are delays in the process. With web-based, virtually there is no installation (if using an ASP model) or very little.

With CST, there could be possible delays. You must wait for IT to install the software into each of the network computers. If the IT department is backed up with other projects or if the IT department does not see maintenance as a priority, installing the CMMS software might be postponed for weeks. If you use the vendor to install the software, it adds to the cost.

Maintenance cost. With CST, maintenance cost is much higher as you have to support multiple PCs. Something can go wrong with CMMS installed on any one of the PCs.

Hardware cost. With CST option, you may have to invest in hardware if current hardware does not support the CMMS software requirements. With web-based, as long as you have internet access, you need not upgrade your hardware.

These lower overall costs make web-based CMMS software a better overall value.

Easier upgrading. With CST, you typically receive upgrades every 12 to 18 months. You have to go through the process of installation and training as it is almost like having a new product. If you have multiple sites, you experience further delays and add to the cost.

When using web-based CMMS, you can be certain that your service provider will automatically upgrade the program. This means that you will never have to receive any upgrades in the mail, email, or download format and spend time installing it. With web-based, the upgrading is an on-going process. That is why there is very little training, if any, required.

Lower IT costs. Another benefit of web-based CMMS is lower IT costs. With ASP option, there are no installations, daily backup, and hardware headaches. Less work for a company's IT department means less money spent.

Greater convenience. End users feel comfortable using their web browser; they are already familiar with common browser elements such as links, buttons, and forms.

Less potential for viruses. Web-based CMMS has much less potential for viruses. With web-based, you are running the program off a browser so there is no local data to damage. If your disk crashes due to a virus, nothing to worry about, the web-based server is automatically backing it up at all times.

Higher security. Contrary to the belief, there is a much higher level of security in web-based software. When using web-based CMMS, your service provider usually has an enterprise-class firewall to prevent all unauthorized users. Additionally, when you access the system, data passes back and forth under the highest encryption available in your Internet browser. Incidentally, this is the same type of security that banks use for online transactions.

Another beneficial security feature is confirming data entry against predefined criteria. For example, users who forget their password might be asked a question that only they would know. Only after answering this obscure or personal question, could the employee gain access into the web-based CMMS.

Customer support. With CST, you often hear from vendor's tech support, " Well, I can't really see what's wrong. Can you send your data to me? I'll take a look at exactly what's going on and get back to you."

With web-based CMMS, provider's tech support can log right in and see exactly what you are seeing and then troubleshoot. That means faster resolutions of problems, concerns, or questions.

Multiple facilities (plants). Multiple facilities are much easier to handle with web-based technology. Each facility has its own database. Yet, managers with right passwords have access to other facilities' data. This is a very useful feature. For example, if one of the facilities is out of a critical part, you can see if any of the other facilities has that part in stock. Your problem can be solved in minutes.

Also, multiple locations can work on the same real time data all the time. There is never a need to merge, synchronize or duplicate CMMS data. CMMS reports are always up to the minute, no matter where you are.

Reliability. Most web-based service providers claim an uptime of 99 percent or better which is probably better than most CST-based CMMS. This is because the service provider has an excellent IT staff and data center facility with redundant everything. You get these benefits without having to pay for all the hardware and staff yourself.

Data entry. With CST, data entry is always within the network. For example, you cannot enter data from home or if you are out of town. If you did, there will be a problem with synchronizing the data. With web-based, you can enter the data from any where in the world.

Back up. With web-based ASP option, data is backed up continuously. If you like your own back up, besides their continuous back up, there is usually a provision for that.

Product demonstration. Try before you buy it. It is much easier to obtain a demonstration with a web-based system. You can just go to vendor's Web site where demo is offered, receive a user ID and a password in seconds without any human interaction. You could be viewing the demo in minutes, no software to install. You are not waiting for the package to arrive in the mail either.

Bugs. With web-based, everyone is always using the same version. Bugs can be fixed as soon as they are discovered. There are two types of bugs. First type is serious in nature that has to be fixed immediately. With CST, vendor sends a CD to all users or fixes are sent via email or download. Either way the user has to go through installation. Imagine finding a bug of this nature and having to go through this with hundreds or thousands of customers.

Second type of bugs are not serious in nature, they cause inconvenience however, or are annoying a bit. With CST, you accumulate these kinds of bugs and send the fixes along with another release. Until then the users have to put up with annoyance or inconvenience. Also, with CST, you normally fix a "pool of bugs" at one time closer to the "upgrade release date." Fixing a bug in something written six months ago is a lot harder and a lot more work. With a web-based system, you fix bugs as soon as they are discovered.

Conclusion. In conclusion, the advantages of web-based CMMS are clear. It offers better-quality service at a lower overall cost. The fact that plant and maintenance managers can use web-based CMMS software from anywhere is very beneficial. With better service, lower overall costs, and more convenience, it seems there is only one question remaining. How can any plant or facility afford not to invest in web-based CMMS?

Mobile technology with CMMS.* One of the key factors in the success of a CMMS implementation is the amount and quality of the data that populates the system. Most organizations admit to only using a fraction of their CMMS capabilities because they frankly find it near-to-impossible to collect the vast amounts of data necessary to feed the endless features available to them. The flaw is not in the CMMS themselves, but rather how these systems are populated with data and at the very basic level,

*This section is contributed by Erica LeBorgne, Syclo, Erica.LeBorgne @syclo.com.

woven into a technician's daily workflow. The fact is, many organizations are still using paper as the key means of collecting information about asset performance and maintenance activities. And paperwork creates an unnecessary bottleneck in information flow to and from technicians working at the point of performance. Without the right data, organizations are not gaining the most from their investment in CMMS.

Mobile computing offers a proven way for increasing the quality and quantity of data populating CMMS systems. Many maintenance departments have realized the vast inefficiencies and costs related to a paper-based workflow and have adopted the best practice of using mobile computers to collect data. Rather than sending skilled technicians out in the field with a clipboard, they are given mobile computers instead, with a full list of their work orders, detailed job plans, and asset histories readily at hand.

As most companies begin to deploy mobile technology, they soon realize that the end goal of improving data in their CMMS is just one of the many benefits that they experience. Mobile affects all aspects of the workflow and maintenance processes—from the back-office where administrators and planners are freed to focus on new tasks to the field where technicians become more productive and effective to the manager's office where an accurate and timely look at asset performance finally becomes attainable.

Equipped with mobile computers—*personal digital assistance* (PDAs), tablet PCs, rugged laptops—and appropriate software, technicians can quickly and easily access the data they need to better troubleshoot. Customers have reported productivity increases of up to 28 percent. First-time fix rates generally increase. Inventory management becomes more efficient with better tracking of spare parts. Planning and reporting for compliance also become easier with the influx of data supported by mobile. Advanced mobile technologies such as *radio frequency identification* (RFID), *global positioning systems* (GPS), *geographic information systems* (GIS), and digital image capture also make it easier than ever for maintenance departments to support the needs of their mobile work teams and capture the critical data that is necessary for optimum asset performance.

Data captured in the field using mobile technology is more precise than information previously captured on a piece of paper. With mobile, data that is entered must conform to the standards required by the software application. For instance, administrators can set limits for readings entered for particular assets. If a technician enters a measurement that exceeds the allotted range, then an alert may require they reenter the information or confirm the reading.

The more data points a CMMS has to work from, and the more accurate the information, the better the reports it generates.

The benefits of mobile computing are numerous, and will be discussed in detail later in this chapter. For now, let us examine the technology behind mobile computing that makes all of these valuable benefits available.

The technology. Before further discussing the business case for mobile technology, it will help to have an overview of the technology's key components (in layperson's terms) and define what "going mobile" really means. The three main categories to consider are mobile devices, software, and data communication.

Mobile devices. Mobile devices are a primary component of the application. These are either PDAs (i.e., Pocket PC, RIM Blackberry, Palm One Treo, and the like), tablet PCs, which you can think of as digital clipboards with computer screens, or laptops which can come protected in layers of heat-, water-, and shock-resistant plastic.

PDAs can hold megabytes of data, including catalogs, spec sheets, WOs, maintenance histories, and work forms. Data can be input into or accessed from the device using a keypad or penlike stylus. Some devices have color screens and can be armed with bar-code scanners or RFID readers.

Average PDAs are relatively inexpensive, ranging in the hundreds, but ruggedized PDAs cost a bit more, in the thousands and have a longer life. Tablets and laptops have bigger display screens and hold much more data, but are heavier and cost more and also pose greater limitations in terms of portability and battery life.

These devices add value to CMMS applications because technicians can (1) carry more information than is possible with paper-based systems, (2) complete WOs faster and more accurately, and (3) get data that's collected in the field back into CMMS faster. The PDAs are lighter and easier to carry, but some work situations may demand tablets and laptops that have more computing power and bigger screens.

A crew assigned mainly to break/fix tasks may use rugged handheld devices with wireless communication so they can report each completed trouble ticket and receive new work as assigned. Inspectors may use a tablet because more data needs to be stored on the device and viewable on screen such as past inspections or out-of-range limits data. Managers may track work using laptops mounted in their vehicles. The key to device selection is to have a choice, plus the ability to mix and match devices within an organization.

Software applications. Mobile application software enables developers to create electronic forms, transforming key data from CMMS applications into usable WOs that make it easier to view on devices, and create an easy-to-follow workflow. Mobile software will be integrated directly with existing CMMS and will enable end users to pull data

Figure 2.27 Mobile technology in use.

directly from the backend system and update information remotely (See Fig. 2.27). Mobile software also manages the synchronization of data between the devices and application database.

Where significant *return on investment* (ROI) is realized is where the software does one of the things that paper forms can never do—enforce business rules. Managers can send mandates to the field from the office, but only software can force a technician on-site to enter a failure code when completing a break/fix work ticket or specify which parts were used to repair an asset before closing out a work ticket.

Many tasks, such as rounds or inspections, require that the software facilitate data collection and transfer. But technicians may be required to perform other tasks once they have mobile devices such as create work requests on-site or complete a PM task that was forthcoming for the same asset they are inspecting. The mobile software should make this PM appear to the worker and require its completion.

These days, mobile software made for organizations with small IT staffs and/or a workforce, which is moderately technology-savvy, is typically designed to be easy to install and learn by technicians. The cost of a mobile software application typically includes the time required to deploy it, train technicians to use it, maintain the software and upgrade it.

For optimal integration with CMMS applications, mobile software needs to be 100 percent configurable so data presented on the screen is in your workers' familiar vernacular and matches their existing work processes. This generates faster acceptance and greater usage than presenting data the way it appears on the CMMS desktop computer screen.

Wireless access. Though mobile devices work without wired connections to network servers and to the Internet, they do not have to use cellular phone carriers' wireless networks to transmit data. Data transfers can happen by placing the devices in a cradle that links to a desktop or laptop. Data can be transferred back and forth between the device and the computer that in turn is connected to your network servers.

That said, there are some advantages of using a wireless connection to transport data. Break/fix people can immediately receive new WOs and WO changes after they have started their rounds regardless of where they are. Data that inspectors collect in the field may have a greater impact on CMMS data analysis processes if it is immediately uploaded through this connection.

In both cases wireless cellular network access, often called wide-area wireless, enables these benefits. Wireless carrier charges for this access time are usually more than double the cost of cell phone time. Many organizations control costs by having their workers only connect occasionally and just long enough to exchange data. Workers can store, access, and perform most computing tasks on mobile devices without needing to be connected.

An alterative to wireless carriers is the use of wireless LANs (commonly referred to as WiFi or hotspots). Throughout a plant or office campus, access points with wired connections to a LAN relay data wirelessly from mobile devices and on to LAN. During an inspection or maintenance job, workers' devices can access data from the CMMS, or upload collected data. This wireless exchange is free if you own the access points and LAN.

Ideally, mobile devices that can switch easily between "always on" wide-area wireless, WiFi and cradle data syncing without losing data in the process are the best options. You can control costs, communication efficiency improves, field workers can operate most effectively, and the office staff can manage both the data and the workers better.

The real cost of paperwork. Paperwork is often seen as a necessary evil, an inefficient business practice that many organizations have become so accustomed to that they no longer view it as a problem that can be solved. But paperwork soaks up even more energy and resources than we realize, and creates an information bottleneck that decreases the potential benefit a manager could gain from being able to view real-time assignments and status of WOs.

A leading pharmaceutical company found that a paper WO had a life cycle of eight key steps—passing from a planner who generates, assigns and distributes WOs to the technician who notates work completed as he performs his job, and must later complete additional paperwork to be turned into a data entry designee. That data entry designee would then have to collect and review WOs for completion before entering data into the EAM system and filing the paperwork.

The total cost to complete a single paper WO was calculated to be $4.90. The pharmaceutical company conducted a comparative assessment of the new 4-step process required in completing a WO using mobile technology, and found the average cost per WO was $.84. This was a compilation of basic administrative overhead and a technician's lost productive time, not factoring in the benefit to the overall maintenance operation when timely access to WO status and data was made available to supervisors for better reporting and planning. (See Figs. 2.28 and 2.29)

How can you quantify the effort of your current work processes? First define the "life cycle" of a WO from the creation of the WO; through the assignment and management of an individual job; to the communication methods used to distribute WO information; to final reporting, whether that includes only a paper hand-off or a combination of a call back, fax, data entry, and so forth.

Next consider the steps that are required specifically of the technician and which of those steps can be eliminated such as trips to a computer terminal to enter updates after each WO, or the paper hand-off at the start of shift. Finally, quantify the effort of the current work process including the time to generate, distribute, and report on WOs, and note the impact of eliminating entire steps. Not only will you gain in the discrete time to actually complete the activity, but you will also eliminate excess foot traffic and breaks in productivity that might revolve around socializing, taking a "break," meandering, and the like.

Just to get a small sense of how these inefficiencies translate into a financial impact on your business, take a pen and paper to calculate just one area of ROI shortfall.

- How much does the average technician cost you per hour (salary, benefits, and so on)?
- How much does an administrator who performs data entry cost per hour? How much time do they spend trying to contact technicians to clarify illegible entries?

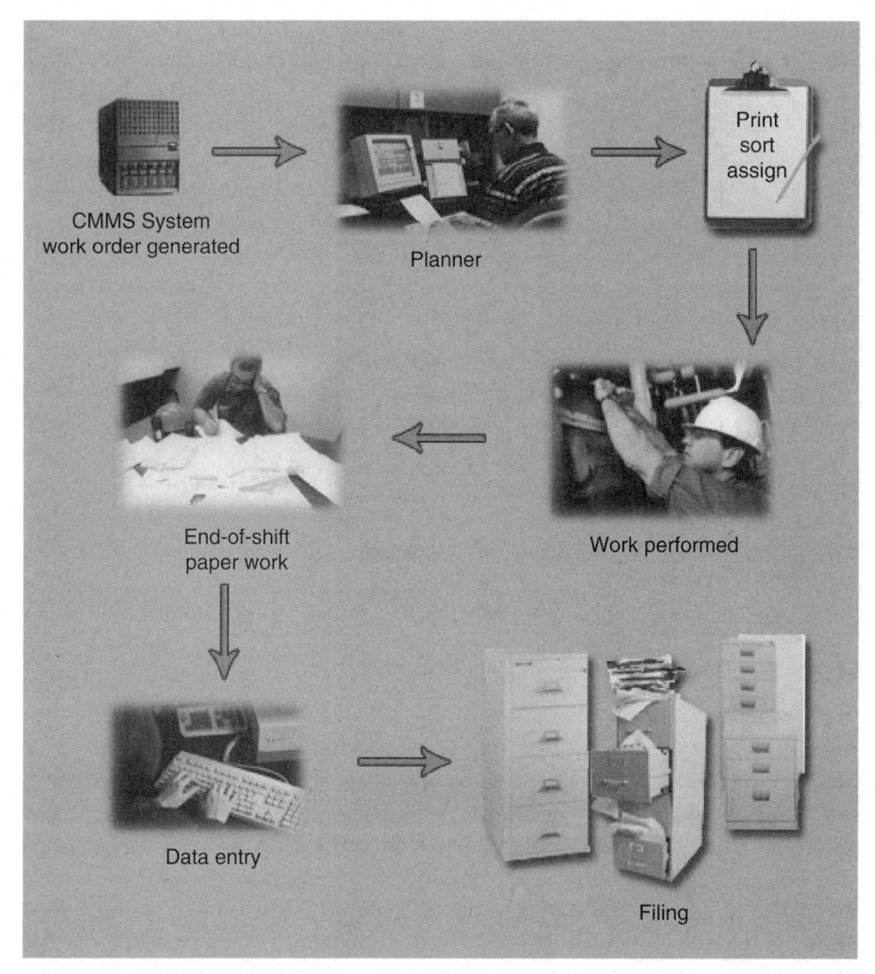

Figure 2.28 Paper workflow.

- On an average day, how much time does a tech spend completing paperwork? Multiply this by a tech's hourly cost. Multiply the hourly cost doing paperwork by the total number of technicians. This final number is your *daily* burden of a paper-based system.

To zero in on these costs, mentally follow the average worker for a day. What are all the little costs that add up to big money in a process that leads to limited data collection, creates errors in reporting time on a job, and is prone to causing workers to miss critical data? Typically only 30 to 40 percent of the required information is captured in the CMMS.

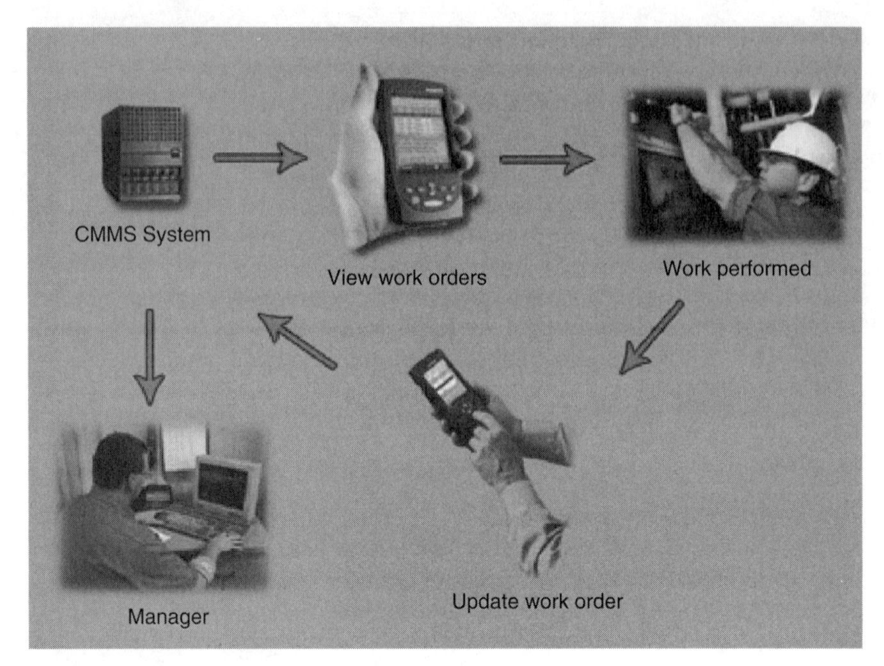

Figure 2.29 Mobile workflow.

Mobile with CMMS. So what can mobile offer the maintenance world in terms of workforce automation? Mobile computing fundamentally changes the way technicians interact with their backend maintenance or asset management system. Instead of a paper-based workflow, where interaction with the system may only be session-based, as in when making updates, mobile provides the opportunity for workers to be highly interactive—constantly in and out of the system.

This drastic change in how maintenance technicians now interact with a CMMS does not mean data entry responsibilities have merely shifted from the clerk and out to the more expensive resource, the technician. Rather, mobile changes how the application is used as a problem-solving tool and a means of providing managers with timely, accurate visibility into actual work being performed on assets. The application moves from the back office and is mapped onto mobile devices so that it becomes an integral and natural part of doing a technician's job.

Benefits of mobile technology. The benefits of deploying CMMS are expanded to include a new subset of benefits that provide greater payback on the initial investment, including increased productivity, more/better data that allows for better planning and reporting, and an increase in technician success with access to data at the point of activity, which helps improve first-time fix rates. Over time, mobile has helped maintenance organizations increase asset touches, reduce maintenance backlogs, increase predictive maintenance, and extend the overall productive life of critical assets.

Eliminating paperwork reaps instant productivity gains. The most tangible of mobile benefits starts with eliminating data entry and paperwork. The cost savings can be found in the elimination of administrative overhead and a gain in wrench time once technicians no longer need to be concerned about end-of-shift paperwork sessions, which last an average 43 min per shift. Furthermore, information at the point of activity helps improve productivity by giving technicians the data and answers they need, without contributing to excess foot traffic in search of asset history, and the like. Finally, mobile helps increase the quantity and quality of data for decision support—in essence feeding your CMMS.

The value of information at the point of activity. Beyond the inefficiencies in workflow created by paper, consider the limitations of a static piece of paper. Paper offers no interactivity to be gained from technicians who need access to specific information in order to complete their jobs more quickly and effectively . If data is missing from a paper WO, your technician must go in search of the information he or she needs to complete that job.

Consider that your organization purchased an asset management system to improve information flow in your organization, yet you have not even started to reap the benefits of such a system until you devise a way to effectively communicate vital information to the people who can most benefit from it. To work smarter, technicians need more access to information—information contained in the CMMS.

Currently, unplanned paper WOs usually only provide technicians with the name and location of the asset to be repaired and hopefully a description of the problem. Planned WOs typically contain some more details, including a description of the PM to be performed, the location of the equipment, the job plans, and maybe even warranty information. Yet the information is dated. With mobile access to their CMMS, technicians can access the history of a WO, measurement history and action limits for a specific piece of equipment, warranty information, safety and job plans, bill of materials, parts locations, failure history, and accurate equipment history—all up-to-date.

Quantify the value of having information in technicians' hands by looking at the potential results. With WO history, first-time fix rates improve. With parts information and availability, foot traffic is reduced, as is the time to complete a repair. Failures can be reduced with condition monitoring action limits available from a handheld device. Compliance and safety improvements can be made with access to job/safety plans at the point of performance. Duplicate work is eliminated because work will never be assigned twice, as is common when completed WOs were late in being entered into the CMMS system. Finally, up-to-date information can improve collaboration between technicians.

More, better data for management: planning, scheduling, and reporting. As stated earlier, a CMMS is only as powerful as the data in the system. Managers are left virtually blindfolded unless CMMS data is kept accurate and up-to-date; their ability to make sound planning decisions is impaired. Effective maintenance strategies including *reliability centered maintenance* are 100 percent dependent on accurate and timely feedback and recording of asset touches. Detailed labor data is necessary in order to determine where to use your limited human resources. There is always more work to do than people to do it. Having the accurate and timely information at hand is what allows managers to direct their people wisely to where they will have the most impact on improving compliance, customer service, or production.

Another example of data accuracy gained with mobile is the usage of failure codes. CMMS require failure codes for root-cause analysis, yet technicians may forget to enter a failure code or may not have standard failure codes that they use in a paper-driven work environment. One technician's "leaky pipe" could be another's "faulty plumbing." Mobile software can require technicians to select a failure code from a list of predetermined codes. This is a critical piece that maintenance organizations require in order to benefit from their CMMS.

Another increasingly urgent issue maintenance organizations face is coming from regulatory agencies, which have stated and demonstrated that work that is not reported has not been completed from a compliance perspective. In regulated industries, if maintenance managers cannot instantly run a report and prove that certain assets have been maintained or repaired, there will almost certainly be penalties or fines to deal with. Paper-based workflow cannot protect against regulator's requirements like a mobile solution that includes e-signature or e-verification capabilities.

Mobile solutions for CMMS have been deployed within hundreds of companies worldwide, across industries. Please see App. 2D, "Mobile Technology Case Studies."

Conclusion

Moving workflow from "work, then record" to "record as you work." While understanding the high-level benefits of mobile, organizations may still have concerns that end-user adoption may be slow and impair the prospective gain in productivity. There is a fundamental change that takes place when replacing a paper WO with a mobile device-one that can be referred as "work, then record" versus. "record as you work." With paper-based workflow, technicians often sort their WOs, complete work, and then set aside time at the end of the day to complete paperwork. With mobile, technicians are required to record information as they work, to close out one WO and move onto another.

There are some clear advantages to mobile providing interaction with the backend CMMS system, which would rarely or never occur when technicians are tied to static, paper WOs. The fact that the mobile application is interactive means it can require users to enter necessary information; it can prompt action to be taken when measurements are within certain levels; and it can also allow technicians to proactively initiate a WO.

As stated earlier, important fields such as failure codes may be required to be entered before a WO can be closed, ending the all-too-common practice of entering failure codes based on "best recollections" or a simple "no trouble found." With more accurate labor figures, failure codes, and parts usage being entered in a timely manner into the CMMS, managers can actually begin to develop maintenance strategies and manage inventory around failure information that is meaningful.

These nuances in information go a long way in improving a maintenance organization's overall ability to prevent and detect failures. If actions are performed as part of cyclical PMs, managers can also use this data to predict quarterly supplies needed to better manage inventory costs.

With more accurate labor figures, recording of work being performed, and parts usage being entered in a timely manner into the CMMS, managers have greater visibility into day-to-day operations for better planning and regulatory compliance. They can also begin to develop maintenance strategies driven by meaningful equipment failure information and a 360° view of asset behavior.

At this point, mobile moves from solely making workflow more efficient to the greater purpose of becoming an enabler for developing asset management strategies driven by accurate and meaningful data points. As increases in productivity eliminate work backlogs, PM measures begin to overtake corrective maintenance.

Without the constant fire drill of break-fix dispatching, and with more information available on assets, managers can begin taking proactive measures to schedule and perform maintenance before a valuable piece of equipment exceeds its performance threshold, resulting in costly downtime. Condition monitoring conducted by technicians who upload findings into a CMMS in real-time provides managers the upper hand in making critical decisions to keep valuable assets working. Replace-versus-maintain decisions are also made easier, with the ability to calculate the actual cost of maintaining a particular piece of equipment.

Optimizing asset performance requires that organizations create an environment where asset maintenance is knowledge-driven. Knowledge and data delivered to and from the point-of-performance can help an organization gradually progress its maintenance strategy to become reliability-centered. To get to that point, however, CMMS must be populated with accurate data in a timely manner. Mobile is the key driver in helping organizations reach this goal.

Automatic identification and data capture for maintenance management. Like any other mission critical activity, maintenance management is driven by information. Computer systems automate many aspects of a maintenance operation, usually relying on keyboard input and paper output to collect and disseminate information. Maintenance operations cannot avoid the computer keyboard and paper report. However, there are situations where the entry and publication of maintenance data can be automated. Given the right situation and proper implementation, this automation can significantly enhance the effectiveness of a maintenance operation.

The process of automating the entry and dissemination of computer-based information is called *automatic identification and data collection* (AIDC). According to the *Association for Identification and Mobility* (AIM), AIDC "is the industry term which describes the identification and/or direct collection of data into a microprocessor controlled device such as a computer system or a *programmable logic controller* (PLC), without the use of a keyboard." Based on this definition, one might believe that AIDC solutions are only deployed in the back office or automated control systems with little or no direct impact on our daily lives and workplace duties.

Actually, AIDC is quite commonplace in our society. Most people directly interact with AIDC-based solutions every day. We encounter them at the grocery checkout line. We use them to pay tolls, buy groceries, and get cash from ATMs. In the workplace we employ AIDC to route letters and parcels, access buildings, manage inventory, track documents, prevent theft, drive point-of-sales systems, and enhance the safety of food and pharmaceuticals.

But how often do maintenance organizations use AIDC to directly support their operations? If we discount magnetic striped credit cards used for procurement or the RFID-encoded building security cards, most maintenance departments do not use AIDC solutions to support their daily work functions. In many operations, maintenance is more likely to encounter AIDC supporting other enterprise functions than embedded in maintenance systems.

Can AIDC improve the performance and efficiency of maintenance systems and processes? Like any technology used in the workplace, the answer depends on a variety of factors. Given the right situation, an AIDC solution can produce substantial ROI for a maintenance operation. But not every maintenance department can benefit from employing an AIDC technology. Deciding if AIDC is the right approach for any particular situation requires a basic knowledge of how the technology works as well as solid understanding of its benefits and challenges.

AIDC encompasses an assortment of technologies that provide machine-based alternatives to keyboard data entry. These technologies include:

- Bar Codes
- Radio frequency identification (RFID)
- Contact memory
- Voice recognition
- Biometrics
- Magnetic stripe cards

The definition and boundaries of AIDC are not rigid. It is frequently called by other names such as *automatic data capture* (ADC) or *automatic identification technologies* (AIT). Other technologies share similar characteristics and benefits, but are typically not categorized as AIDC solutions. Condition monitoring equipment can be used to capture run time information without direct keyboard entry. GPS technology can allow systems to track global container movement throughout the supply chain. But these technologies are generally not included within the AIDC family.

Each AIDC technology has its own set of unique capabilities and a cost threshold that can make it appropriate for some applications and not others. Some applications use a combination of the technologies, while others can only be addressed by a particular technology. AIDC maintenance management applications are not restricted to the technologies listed earlier. New technologies will also continue to emerge. However, the ones listed earlier represent the current focus of most AIDC activity in maintenance management.

The technology

Common elements in AIDC applications. AIDC maintenance management applications generally have four common elements regardless of the technology used. They are:

- Collection medium and encoding pattern
- Reading and writing devices
- Terminals and data communication
- Application software

The collection medium is the physical vehicle for storing and transmitting information. Bar codes, touch memory buttons, RFID tags, and speech are collection mediums. Information is encoded or stored on medium as an intelligent pattern. This pattern can be digital settings on an integrated circuit chip, or the sequencing and spacing of wide and narrow bars on a bar code.

Reading and writing devices are used to store and retrieve data on collection medium. They usually contain firmware allowing them to encode or decode the patterns used to store information. Some devices like a bar-code scanner or a bar-code printer function either as a reader or a writer. Other devices like a RFID transponder/reader can perform both roles.

Terminals provide a mechanism for users to interact with the collection process and application software. Terminals can be categorized by their data communication method. Fixed terminals communicate with a computer system through cabling and wires. Batch terminals are portable requiring users to physically place the terminal in a cradle or docking station in order to upload and download data from the target computer system.

Wireless or *radio frequency* (RF) terminals also provide portability, but allow users to send and receive on a real-time basis. Terminals vary in processing power from simple storage devices to portable computers complete with keyboard and display. Some intelligent terminals are only programmable in the vendor's proprietary language. Others are basically portable, PCs running under Windows CE. Mobile devices vary in size and shape from a credit card to a notebook format. They can come in hand held, vehicle mounted, wearable, tablet, or notebook configurations. They generally run special mobile computing module software supplied by a CMMS vendor. However, other software such as predictive maintenance analysis and stand-alone inventory control packages can support AIDC.

Commercially available software packages do not inherently support AIDC. Vendors must design and develop a special program code in their products in order to support it. IT departments and system integrators can custom build stand-alone AIDC solutions or in certain instances, integrate AIDC technology into an existing application package.

AIDC with CMMS.* AIDC can be employed to support a variety of maintenance activities. The MRO storeroom is a popular setting for AIDC where it is used to support issues, receipts, and inventory counts. It can also secure storeroom access and help support self-service operations. Other application areas include:

- *Corrective and PM WO.* Record key activity and service data, and support mobile applications.

---

*This section is contributed by Thomas Singer, Tompkins Associates, tsinger@tompkinsinc.com.

- *Asset tracking.* Capture asset/equipment identifiers and track items by location.
- *Inspections.* Record and store inspection data.
- *Time and attendance.* Capture employee identification and activity codes.
- *Security.* Control access to buildings, equipment, and vehicles.
- *Tools and key tracking.* Support check-in/check-out, track location, and record warranty and service information.
- *Condition monitoring.* Support asset identification, and data capture and collection.
- *Fleet management.* Support fuel usage collection, record service data, and track location movement.

EAM and CMMS vendors are the primary sources for AIDC maintenance management solutions. This is not surprising since they provide the basis for most maintenance management information needs. Bar coding is the most popular technology supported. See Fig. 2.30. Many EAM/CMMS vendors provide bar-code modules or support the printing of bar codes on maintenance documents. Most EAM/CMMS mobile computing modules either directly embrace AIDC or can become AIDC-enabled through the addition of a bar-code scanner, touch memory probe, or RFID reader to the handheld device.

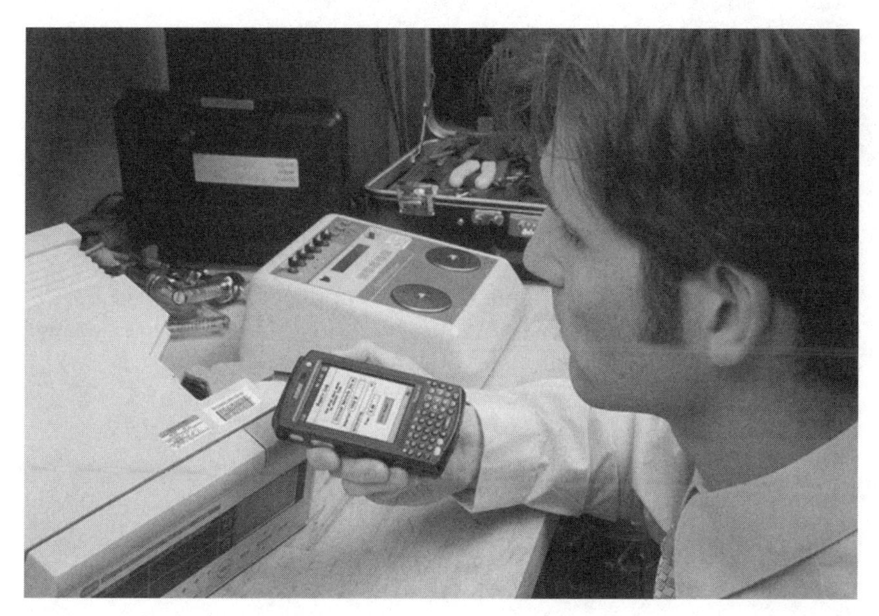

Figure 2.30 Bar-code technology in use.

Specialty solution suppliers are another source of AIDC maintenance management solutions. These vendors provide standalone applications to support specific work functions and activities. Some offer AIDC "bolt-on" solutions that can be integrated into legacy enterprise and EAM/CMMS systems. Custom development is also potential source for AIDC applications. Regardless of source, these solutions can cost hundreds of thousands of dollars to deploy or they can simply entail the acquisition of some relatively inexpensive hardware and PC-based software.

Benefits of AIDC. AIDC can benefit a maintenance organization by:

- *Reducing the time spent on data entry.* Data captured through an AIDC technology saves the labor time that it would take to key enter it.
- *Increasing the accuracy of maintenance information.* Even the best data-entry operator makes mistakes and maintenance personnel are typically not hired for their data-entry skills. AIDC can reduce the incidence of human error in the data-entry process.
- *Reducing paperwork.* Paper is cumbersome and costly to process. AIDC can help streamline the dissemination of information.
- *Identifying assets.* AIDC can make asset management easier and more reliable by providing a machine-readable serial number.
- *Supplying information where it is needed.* AIDC can make maintenance information mobile by providing it to maintenance personnel at the job site.
- *Providing an activity audit trail.* AIDC can help verify that critical maintenance activities are occurring by recording the arrival of personnel at the job site.
- *Securing valuable resources.* AIDC can control access to the maintenance storeroom or track the dispensing of tools.

Bar codes. Bar coding is an accepted, if not common, practice in maintenance management. There are bar-coding solutions for just about every maintenance system application. Any application that requires the entry of a predetermined set of values such as WO numbers or failure codes is a candidate for bar coding. Bar codes can support:

- WO processing
- Inventory control
- Inspections
- Tool tracking
- Asset management
- Labor reporting

A bar code is a pattern of alternating dark stripes and light spaces. It allows key data elements such as WO numbers, part numbers, and failure codes to be encoded on a piece of paper or label. An optical scanning device reads the bar code by illuminating the pattern and translating the resulting reflection into a data stream.

Symbologies and data structures. There are different methods or *symbologies* for encoding data in bar codes. The type of symbology used is generally dependent on the industry or application. For example, the retail industry uses the *universal product code* (UPC) for product identification. The most popular symbologies used in maintenance management applications are UPC, code 39, interleaved 2 of 5 (ITF), and code 128 (Fig. 2.31). Most scanning equipment and bar-code printers support all the popular symbologies. The choice of symbology in a maintenance application is usually based on the type of data being encoded and its size or field width.

Linear symbologies, like Code 39, ITF, and Code 128, encode information in a single row of alternating white and black stripes. Stripes or bars can either be wide or narrow. Each symbology employs its own bar sequencing and spacing scheme to encode information. Some symbologies like UPC and ITF can encode numbers only. Others support a limited or full set of alphanumeric characters. Code 39 supports encoding of a 43 alphanumeric character set; letters "A-Z," digits "0-9," and a limited set of special characters. Code 128 supports encoding of a full ASCII 128 character set.

Some linear symbologies like UPC also require a specific data structure or format. Retail *point of sales* (POS) systems are based on the premise that each product sold has a unique identifier regardless of its source or original supplier. A 12-digit UPC-A supports this premise through a multipart data structure. The first digit identifies the product or manufacturer type. The next five digits identify the manufacturer or supplier company. These company codes are assigned by GS1 US (formerly known as Uniform Code Council). The second set of five digits identifies the specific product. These product identifiers are assigned by the manufacturer or supplier company. The twelfth position is employed as a check digit used to verify the proper encoding of an individual UPC.

UPC-A is actually a subset of a 13-digit global product identifier known as EAN-13. GS1 (formed by the merger of UCC and EAN

Figure 2.31 Code 128.

International) administers the EAN system. The first two digits of an EAN-13 identify the country issuing the subsequent manufacturer or supplier company code.

Some industry specific data structures utilize certain symbologies. For example, a *general trade identification number*-14 (GTIN-14) is similar in structure to a UPC except it also employs a prefix that specifies the product packaging configuration (e.g., individual units, cases packed with six units, cases packed with eight units). GTINs are typically encoded as ITF bar codes on outer-pack cases.

Most linear symbologies are of variable length. However, there is a practical limit to how many characters they can encode. Their primary purpose is to uniquely identify objects. Stacked and high-density symbologies can encode larger data sets by stacking bar patterns in two-dimensional arrays. Data Matrix and PDF-417 are two popular high-density symbologies capable of sorting up to 3116 and 2725 characters respectively. Data matrix and PDF-417 can be used to encode service instructions, shipping manifests, warranty information, or material safety data sheets. See Fig. 2.32.

Linear symbologies may also produce bar codes that are too long or big to be used on small parts and products. *Reduced space symbology* (RSS) employs compression and stacking mechanism capable of producing a bar code with a relatively small footprint.

Printers. An outside vendor or an on-site printer can produce bar codes. Outside vendors are capable of producing a higher quality, longer-lasting bar-code label than the printers typically found in maintenance departments. Outside vendors are occasionally used by maintenance operations to produce storeroom bin labels and asset tags.

Most bar codes used by maintenance applications are printed on-site. The most popular device for printing these bar codes is the laser printer. CMMS packages that support bar coding typically print the bar code of key data elements directly on their WO forms, material pick tickets, inventory count sheets, and equipment catalogs. Items that must be individually labeled such as parts and storeroom bins are accommodated by special label forms.

Impact, thermal, and inkjet printers are also used to print bar codes. Impact printers are typically used in high volume, batch operations where labels and forms are printed in large bursts. Thermal printers are popular in shipping and warehousing applications where specialized

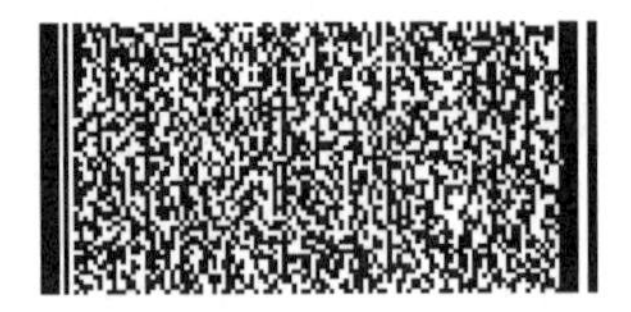

Figure 2.32 High density bar code.

labels with text and bar codes are applied to cases and pallets. Thermal printers work by either applying content to label stock by heat transfer of ink from a ribbon (thermal transfer) or directly heating special label stock (thermal direct). Even though it requires additional ribbon stock, thermal transfer is popular in desktop printers and automated print applicators because the resulting label quality has a much lower sensitivity to heat and direct sunlight than direct thermal labels. Mobile thermal label printers typically employ direct technology because it allows them to dispense with ribbon stock. See Fig. 2.33.

Inkjet printers are also frequently used to print bar codes. Desktop inkjet printers can function in a similar manner to tabletop laser printers. Furthermore, inkjet-printing devices can be used to apply bar codes to corrugated cases as they come off of a packaging line or certain product as it comes off of a production line.

Character density or dots per inch directly impacts the readability of a bar code. Generally, the lower the density the harder the bar code is to read. Bar-code quality is not inherently a function of the print technology used. High-end inkjets can produce high-quality bar codes, while low-end desktop inkjet printers can produce bar codes that are difficult to scan or are unreadable.

Printers need special application software and fonts to produce bar codes. Windows TrueType can be purchased for certain symbologies that allow bar codes to be directly printed from MS-Office applications. Bar-code labeling software is available from various vendors that allow users to custom design label formats that include characters, bar codes, and graphics. These packages typically have the ability to connect with most popular databases. Many EAM/CMMS packages provide the ability to directly print bar codes on WO and pick lists.

Figure 2.33 Zebra thermal printer.

Scanners and Terminals. Bar-code scanners and terminals come in all different sizes and shapes. They range in cost from well under $100 to $5000, and above. Their performance, functionality, and durability are directly related to their price tag. Scanners are categorized by the technology they employ and by the way they are used. Types of scanners are:

- *Contact or wand.* Contact scanners are typically contained in penlike devices that are physically passed over the bar code. The device emits a light that is reflected by the bar code. Pattern changes in the bars are detected by a photo diode. Contact scanners are relatively inexpensive but take more effort and time to read bar codes than other scanning technologies. They also require the user to physically touch the bar-code label.
- *Laser.* Laser scanners operate in a similar manner to contact devices except that they use moving mirrors or prisms to past a laser beam over a bar code. Distance from scan head to bar code varies between models. Some laser scanners operate only within a few feet of the bar code, while others can scan a bar code over 10 ft away. Laser scanner prices have declined considerably in recent years making them the most popular scanning device in the work place.
- *Charged couple device (CCD).* CCD scanners use a linear array of sensors to detect the ambient light reflected off of a bar code. In many ways they operate as a one-dimensional camera that decodes the pattern inherent in a linear bar code.
- *Image scanners.* Image scanners employ two-dimensional array of sensors to capture a complete image of a label. Pattern recognition firmware decodes information stored on the bar code. Unlike laser scanners, image devices can decode barcodes over a wide orientation range. They also handle high density and reduced space symobologies more effectively than other scanner types. However, they are also more expensive. See Fig. 2.34.

Scanners can be connected to a desktop PC or computer terminal through a device called a wedge. This configuration allows a user to scan a data element such as a bar-coded part number on a pick ticket or a repair code on a printed bar-code menu, instead of keying in the data. Given the Windows-based interfaces and sophisticated field-level search capabilities of most modern CMMS packages, wedge scanners have diminished in popularity.

Mobile terminals and computers make many bar-code applications practical by allowing maintenance personnel to capture and process data where the work is being done. Their scanners can be physically integrated into the terminal housing or externally attached to the unit. Many mobile maintenance bar-code applications are batch, requiring the

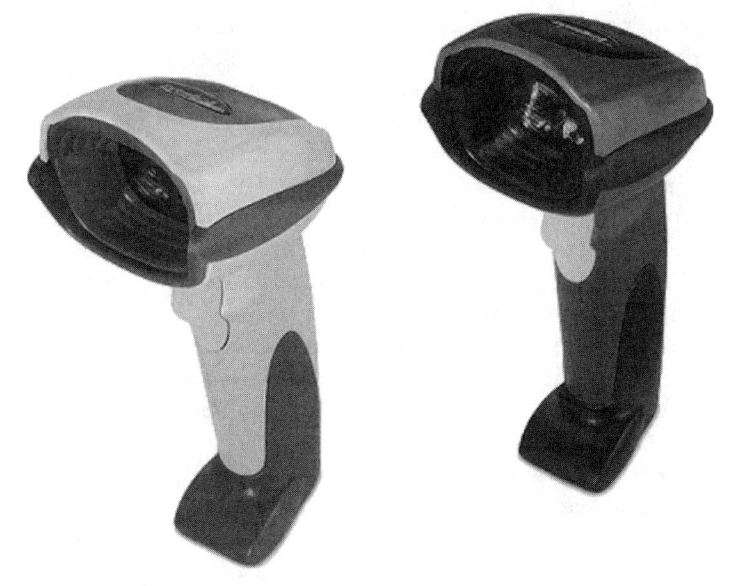

Figure2.34 Symbol scanners.

user to periodically upload and download data from the device using a docking station, infrared port, or wireless connection. Given the increased availability of wireless LANs and cellular WANs, mobile maintenance applications that feature a direct, wireless connect to the host system are becoming increasingly popular. Regardless of whether a mobile application is built around batch or direct data communications, a bar-code scanner can be added to a handheld computer or PDA for a relatively inexpensive price.

Sample bar-code application. One of the most likely candidates for bar-code technology for maintenance operations is the maintenance storeroom. Many CMMS vendors offer bar-code modules that support the maintenance storeroom. These modules generally provide the ability to record WO issues, parts receipts, and inventory counts on a batch or wireless handheld PC.

A typical scenario involves a storeroom attendant picking and issuing parts to a WO. A bar-code material pick list showing all planned or estimated part requirements is printed when the WO is released. The storeroom attendant starts the pick process by scanning the WO bar code. As each part on the pick list is pulled, the attendant scans the part number bar code on the pick list and the location bar code on the bin, and then enters the pick quantity on the portable terminal's keyboard. Unplanned issues are recorded by scanning the part number bar code on the bin or item.

Once the picking process has been completed, the attendant returns the handheld device to its cradle and uploads the pick transactions to the CMMS for processing. The CMMS would reject any transactions that do not comply with its editing rules (e.g., a part was issued to a pending WO or from a storage bin not assigned to the part).

There are countless variations to this scenario. The attendant could scan the part number bar code on both the pick list and bin label to help ensure that the right part was picked. The pick list could be downloaded to the handheld unit. The application could also support direct wireless communications between handheld and the host CMMS allowing the attendant to make stock inquiries on the storeroom floor. This would eliminate the need to upload and download data.

Mobile WO and inspection processing is another popular application for using bar codes. Many CMMS vendors offer mobile WO and inspection modules that run on Windows-CE or Palm OS-based handheld computers. These handhelds can either feature an integrated bar-code scanner or can be configured with an external scanner through an expansion slot. Coupling a bar-code scanner with the mobile device can help craftspeople identify the specific equipment item or asset they are repairing or inspecting.

Given that there may be dozens or hundreds of the same model within a facility, a craftsperson can walk up to a specific item and scan the bar code on its asset tag. The mobile device would then display open WOs associated with the equipment item. It would also allow the users to create a WO for the equipment item. If the craftsperson needed spare parts from an unattended storeroom, he/she could scan the UPC or internal part number bar code to record item usage.

Once again, there are countless variations to this scenario. An inspection module could confirm that the technician actually saw the asset through a bar-code scan. Craftspeople could scan bar-coded menus to quickly identify a failure or repair code.

Biometrics. Biometrics relies on unique personal characteristics to identify individuals. These characteristics can include fingerprints, retinal scans, and face symmetry recognition. The particular characteristic of any person is electronically measured and the resulting pattern stored. Subsequent reads of this characteristic are matched against this key database in order to identify the individual.

Biometrics is very popular in security applications. Cost reductions and hardware improvements have extended its practicality in recent years. Low-cost fingerprint "locks" is available for notebook and desktop computers. Retinal scanning accessories are fairly easy to integrate into security applications.

Biometrics is becoming increasingly popular in time and attendance applications. These applications utilize hand symmetry readers to clock users in and out. They eliminate the need to issue magnetic cards or pass codes to employees. They also eliminate "buddy" punching where coworkers clock each other into the system.

Magnetic stripe. Magnetic stripe technology employs magnetic material typically applied to a credit-card size piece of plastic as the data collection medium. Information is encoded by alternating the polarity of small sections of the stripe. Anyone who has ever used an ATM card at a cash station has used a magnetic stripe application.

Magnetic stripe technology is typically used in maintenance for time and attendance, procurement, and security access applications. When an employee identifier is encoded on a magnetic stripe card, it can be used to control and track access to unmanned storerooms and tool dispensing machines.

Conclusion. AIDC maintenance applications are going to continue to grow in popularity as technology advances and its benefits become more widely known. However, maintenance organizations should carefully consider what their needs are now and in the future. The application of technology is not a goal by itself. It must solve and justify a real need within an organization.

As CMMS vendors introduce more AIDC solutions, maintenance organizations must carefully assess their usefulness to their operations. While there are many AIDC success stories, there are cases where the technology has not produced the anticipated benefits. Some companies have expended considerable efforts on an AIDC project only to abandon it as unproductive or too costly.

AIDC technology is not a panacea. Any maintenance department that believes it can move from a fire fighting to a best-maintenance-practices operation simply by printing bar codes or giving craftsmen PDAs is in for a rude awakening. AIDC technology is not a substitute for good management, competent craftspeople, proper techniques, and appropriate information systems. However, given the right situation and plan, AIDC technology can play an important role in the success of a maintenance operation.

In order to be successful, AIDC or any other information technology cannot be evaluated or implemented in a vacuum. It must be part of an organization-wide effort to achieve maintenance excellence. Potential application of AIDC technology should be part of an overall CMMS needs assessment process. Once the informational requirements and flows of the organization have been established, the suitability of AIDC technology can be evaluated. Functional areas that are prime candidates for AIDC technology, based on its potential benefits, can be identified and incorporated into the CMMS selection criteria. The evaluation of AIDC technology should not stop with the implementation of a CMMS package. Vendors constantly introduce new modules and enhancements. An AIDC module that was not deemed necessary when a package was selected can become a viable solution a few years later. There are situations where custom AIDC solutions can be justified and should be implemented.

The need for AIDC technology is not universal across all maintenance organizations. However, most organizations do need to evaluate its suitability to their operations when developing their CMMS needs assessment. Organizations that are truly interested in pursuing maintenance excellence should constantly look for the right opportunities to apply AIDC technology.

RFID with CMMS*

Introduction. The largest bottleneck that currently exists with any CMMS, is the method or methods that are employed to get data/information into the backend database where it can be utilized by the organization as information to make decisions regarding the operation of the organization. Majority of the time data are entered manually by humans. This process is extremely labor intensive, prone to errors, and is costly for the company and clients of the company.

Methods for automatically capturing field data to provide visibility into an organization's asset status and operations are limited and can be expensive. However, RFID technologies have recently emerged as the leading candidate to provide an effective, cost efficient solution to the data collection problem.

To put the technology into context, RFID is one of many technologies which falls under the AIDC umbrella. AIDC, long used for identifying and tracking items, stands for automatic identification and data capture. This is a term that refers to any system that deploys a method of identifying objects, capturing information about them, and entering it directly into computer systems with little or no human intervention. Bar-code technologies also fall under this classification; however, RFID offers several key advantages over existing bar-code systems. The section will explain the basics of RFID technology and its advantages as it relates to CMMS.

The RFID technology. RFID, like bar codes, is a form of AIDC technology. It is the next generation of asset identification and wireless data capture technology that allows objects to be tracked and identified. This data is then sent to a corporate database or IT system where the information can be used to create business value. The basis of the technology is an electronic tag that contains globally unique identifier (32 to

---

*This section is contributed by Mark J. Gillingham, Cathexis Innovation Inc., mark.gillingham@cathexix.com.

128 bit serial number) and is capable of receiving, storing, and/or transmitting digital information by communicating with a tag reader (interrogator) using radio waves. Potentially, one of the most prolific and pervasive technological revolutions in history, RFID is poised to be beneficial to nearly every sector of business. Its scope of application encompasses many facets of business in virtually every industry. For example, RFID is ideally suited for use in tracking asset maintenance, site access, IT system support, supply chain management, product manufacture, shipping, and retail. These are all facets of a single company, and the use of RFID can address virtually all of their needs. A properly designed system will also allow for a degree of data access, synchronization, and automation—previously unavailable.

Figure 2.35 shows a system diagram detailing how a simple RFID system would be deployed. A typical RFID system consists of three parts:

1. RFID tags
2. RFID readers (also known as interrogators)
3. RFID middleware (management software)

RFID transponders (tags). The RFID tag, ranging in size from a grain of rice to credit card size and larger, is the most important component of any RFID system. There are two main types of RFID tags, *active* and *passive.* The use of active or passive tags is dependant on the type of application and will result in substantially different systems requiring specialized readers and management software for each type.

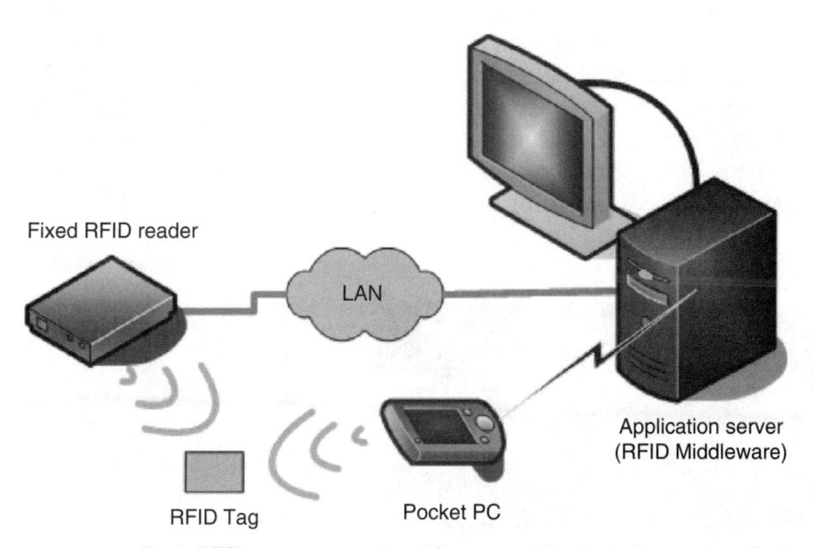

Figure 2.35 Basic RFID system overview. (*Courtesy of Cathexis Innovations Inc.*)

Active tags are usually larger, have great read ranges, are more expensive, and are used in applications for tracking large objects such as cargo containers. The main differentiating feature between active and passive tags is that active tags contain a transmitter and a power source (usually a battery) enabling them to broadcast a signal to a reader.

Passive tags do not contain a transmitter and essentially reflect the radio waves broadcast from the reader. The reflected RF energy contains the unique serial number and any information that may be stored in the tags memory.

Bare RFID tags or inlays as they are commonly known (see Fig. 2.36) are comprised of three components:

1. A microchip that contains a unique serial number and potentially extra memory for additional data.
2. An antenna used to receive and send the tag data to a reader.
3. A substrate (usually plastic), on which the microchip and antenna are mounted.

In most instances RFID used with CMMS will involve passive RFID systems and as such passive technologies will be referenced henceforth. There are many additional technical considerations, which are beyond the scope of this book, to be aware of when selecting a suitable RFID system. The operating frequency (LF, HF, UHF), communication protocol, and the packaging of the RFID tag are examples of several required considerations.

RFID readers (interrogators). RFID readers or interrogators refer to the computing device that is used to communicate with the RFID tag and provide a mechanism to view tag data and/or send the information to a back-end database. RFID readers can come in many shapes and sizes but have two broad classifications: (1) fixed mounted and (2) mobile. See Fig. 2.37.

Fixed mounted readers are intended to be stationary and to be used in processes with mobile tagged assets and are not intended to have a human user. This type of reader is usually mounted in the doorway of a warehouse, directly adjacent to a shipping conveyor, or near the check-

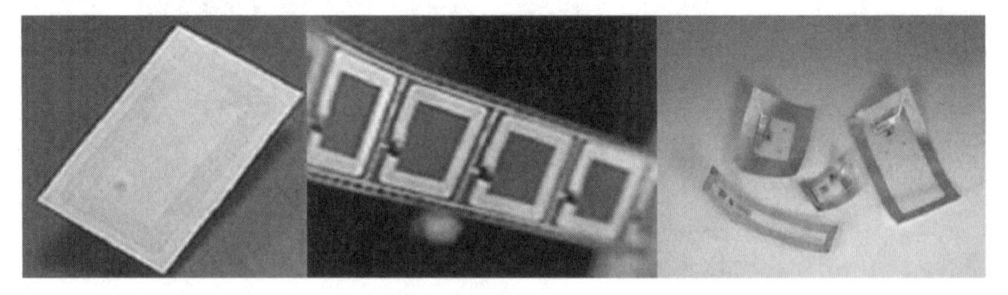

Figure 2.36 RFID tags (transponders). (*Courtesy of Texas Instruments*)

Figure 2.37 IF5 fixed RFID reader. *(Courtesy of Intermec Technologies Inc.)*

out area of a tool shed. The tag data collected by the reader is often sent direct to the backend system via a standard wired or wireless computer network utilizing the RFID middleware management software.

Conversely, mobile RFID readers are usually smaller and more ergonomic than fixed mounted readers and communicate with handheld PCs, laptops, and PCs using RFID middleware management software. The tag data can be viewed locally or sent to a backend system via a standard wired or wireless computing network (e.g., LAN, and WAN). See Fig. 2.38.

RFID middleware (management software). Middleware software, and specifically RFID middleware, is a very widely referenced term and

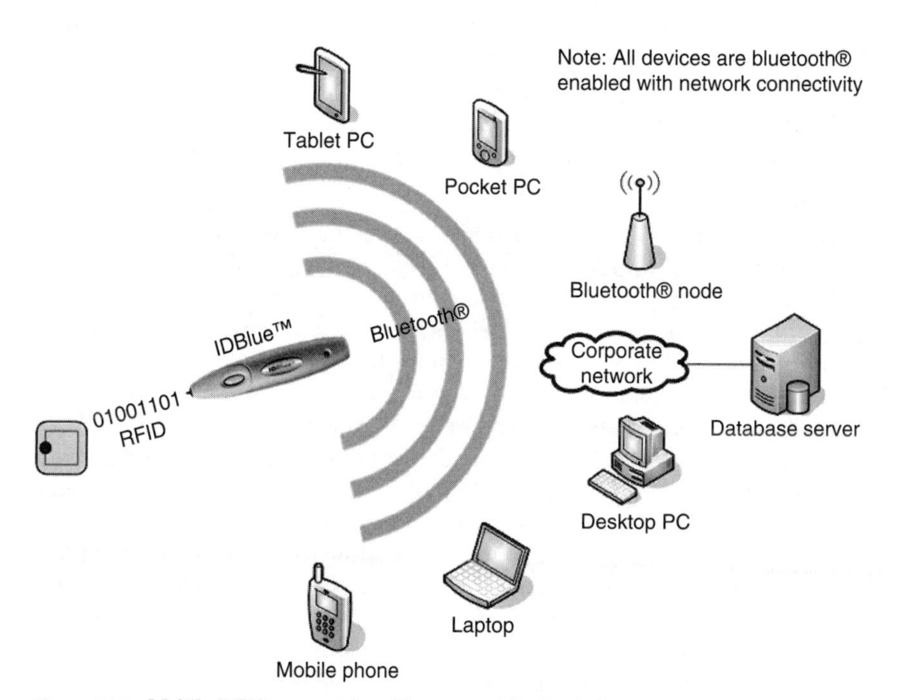

Figure 2.38 Mobile RFID pen reader. *(Courtesy of Cathexis Innovations Inc.)*

is used in many contexts. However, generally speaking it refers to the RFID data management software that enables raw asset data to flow from the RFID tag to a software application (e.g., CMMS) or database. In some instances it is desirable to have data flow bidirectionally (i.e., tag-to-database and database-to-tag).

The main responsibilities of RFID middleware are to manage and monitor RFID data and devices (i.e., readers). RFID middleware encompasses the software that is present on the RFID readers and that facilitates the automated communication and data gathering from RFID tags as well as the transportation of the collected data into backend systems and software applications. Specific functions of RFID middleware include:

- Integration and support of read/write hardware devices (i.e., readers)
- Monitoring status of data and reader devices
- Management of RFID reader networks
- Maintaining security and integrity of RFID data
- Data event management (business rules and processes)
- Data filtering and cleansing

In addition, the software tools to develop RFID-based applications and integrate RFID systems are also classified under the RFID middleware umbrella. The diagram in Fig. 2.39 provides a high-level illustration of how RFID middleware fits between a CMMS software application and the RFID hardware (tags and readers).

Note: This scenario is intended to be a general illustration and as such it may not apply to all CMMS.

RFID uses with CMMS. One of the main reasons RFID technology is proliferating throughout nearly every sector of business is due to its very broad and diverse application set. RFID technology is able to provide significant efficiencies and value to industry-wide problems such as unique asset identification, security, verification and authentication, accurate recording of events, removal of manual data entry, and real-time data flow into enterprise applications—just to name a few.

Each industry has its own, sometimes unique, set of problems where the use of RFID technology can be advantageous. This problem set often gets more specific and complex when dealing with individual client needs and requirements. The goal of this section is not to identify every possible use of RFID with CMMS, but rather to introduce some of the main application areas. Upon finishing this section each reader will probably think of several new ways that RFID can be beneficial to their respective organization and each will certainly be as important and valid as the next. The difficult task then becomes of building a business case and determining what the ROI will be for each use case.

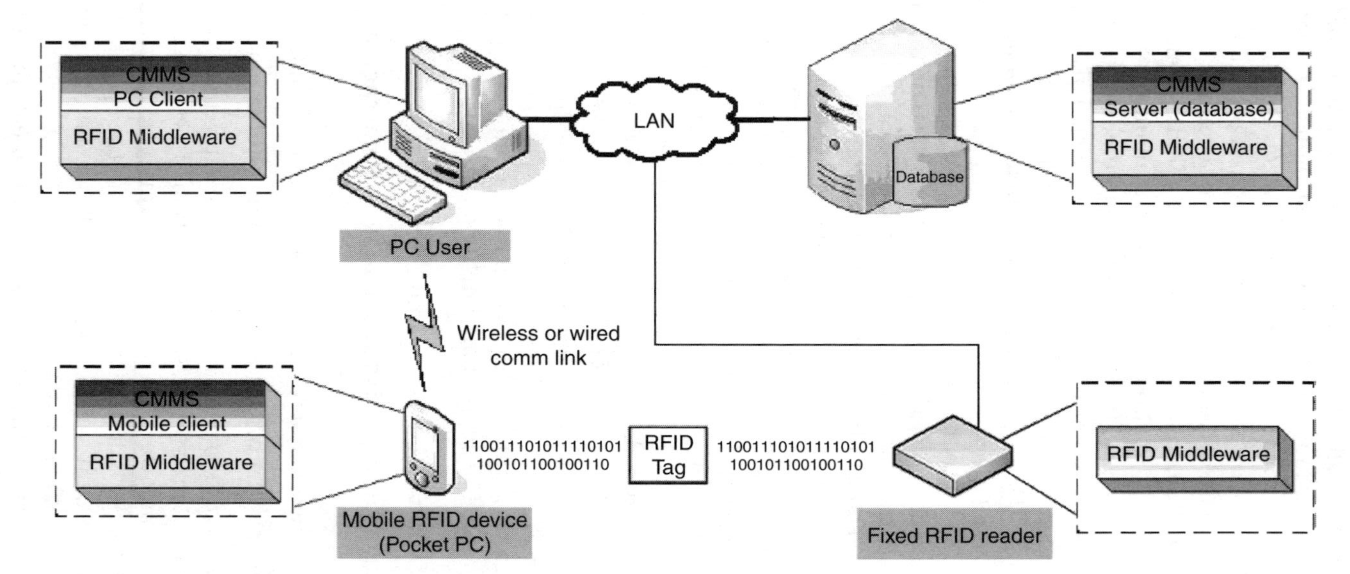

Figure 2.39 CMMS/RFID middleware interface. (*Courtesy of Cathexis Innovations Inc.*)

Unique asset identification. This feature of RFID technology is ultimately what leads to all of the various usages. The ability to identify any type of asset in a unique manner and to distinguish it from numerous other identical items has tremendous application potential.

Ensuring the proper piece of equipment is assigned to the appropriate personnel, or associated with another piece of equipment, or required location can save time, reduce work stoppages, limit resource allocation, and eliminate costly mistakes. For instance, maintenance of aircraft often requires that certain parts, when disassembled, be reassembled with the exact components due to wear patterns. This can be a very tedious and expensive undertaking, but using the unique identification capabilities of RFID, the part associations can be done much more accurately and efficiently.

Automated WO recognition. Once a piece of equipment has been tagged and the unique RFID serial number has been linked to the asset record in the central database, maintenance personnel will no longer be assigned the responsibility of ensuring that they are working on the required piece of equipment. A maintenance supervisor can perform the usual task of generating a WO for a given piece of equipment using the main CMMS. The WO can then be transferred to a laptop or mobile computer (e.g., pocket PC), which is carried with the maintenance workers as they perform their rounds. The system can then be configured to automatically open the WO once the associated RFID tag of a given piece of equipment is scanned.

This application has many benefits especially in cases where there are many similar pieces of equipment or if the maintenance worker is not familiar with the particular location or the specific piece of equipment. Using RFID in this manner can also enable a completely paperless WO system which has the additional benefits of:

- Removal of manual data entry and the associated errors
- Dramatically reduced delays in updating WO information in the main CMMS
- Verification of worker accessing the equipment (with timestamp)
- Eliminating misidentification of assets and equipment

In-field access to information. Another great feature of RFID technology is the ability to store asset information directly on the tag. RFID technology can empower field workers to rapidly access required information and to make educated and informed on the spot decisions without having to consult with other individuals, manuals and printed media, or manufactures and suppliers.

Having incorrect parts installed on a piece of equipment or not following the proper maintenance process is often the result of lack of information. Nameplates often get worn or covered in grease and dirt, and therefore information required to properly complete a maintenance procedure is not always available. The worker is then faced with a situation of having to complete the job without the required information, spend extra time hunting down the required information, or leave the job unfinished and potentially not returning to complete the work.

By having an RFID tag installed on all pieces of equipment, workers can be empowered with immediate access to vital pieces of information that can allow them to do their job much more effectively and efficiently.

The information that is stored on an RFID tag is completely definable by the user (or company) and is often specific to the process or equipment that is involved. This information can be mapped from the existing CMMS data fields to the memory map on the RFID tag. Common types of information that can be stored on the RFID tag are serial numbers, make, model, drawing numbers, inventory levels, maintenance history, and maintenance processes.

Inspection points. Inspection points can cover a broad range of applications, but the main value behind the exercise is to provide validation, verification, and authentication that an event occurred at an exact time. RFID can prove to be tremendously valuable in these instances as unique object identification numbers can be obtained from the RFID tag as well as the generation of an accurate time stamp.

Some common uses of RFID for inspection points occur where equipment is required to be certified or adhere to laws and standards. Such applications include the inspection and certification of fire extinguishers and high-pressure cylinders, emergency exit lighting and doorways, equipment in the oil and gas, or power generation sectors as well as safety inspections on various pieces of vital equipment in the healthcare sector-just to name a few.

Additionally, RFID tags can be used to tag locations that must be attended during security guard rounds. The scanning of an RFID tag can provide secure verification that a guard visited a given location as well as the exact time of the visit.

Identification of facilities. Thus far most of the CMMS applications involving RFID have been centered around equipment and parts. However, RFID tags can be used to uniquely identify anything animate or inanimate. Another great use of RFID is to tag entrances/exits, rooms, or points of interest located within a facility. These tags can be used as information points to help individuals navigate through the facility or to provide information about the contents of a room. Once again the possibilities of using this technology are virtually endless and everyday people are finding more and more creative ways to create efficiencies in their organization by utilizing this great technology.

Emergency data access. The ability to access vital data during times of emergency can prove to be invaluable in helping to get equipment back on-line faster or determine the cause of the problem. Because passive RFID technologies do not require a power source, the information stored on the RFID tags can be retrieved via mobile battery powered computers (with an RFID reader) at virtually any time. Subsequent to instances of power outages, natural disasters, or military style attacks, maintenance technicians and field workers can be isolated from access to information pertaining to essential equipment due to failure of communications networks. During these instances RFID can provide a means of accessing information in-field, directly from the tagged equipment.

Noting that even though these types of instances are rare, rapid access to information at the moment it is required, can potentially save lives and tremendous expenses.

Real-time report generation. Another benefit of RFID is that *all* information that is collected and used is constantly in a digital format. Even information that is updated by field workers via handheld computers is stored in a digital format. This facilitates the rapid and virtually effortless synchronization of field data with the central CMMS database. The benefits are not limited to reduced costs and more accurate data (i.e., no manual data entry and automated data collection), but having real-time reporting and status updates of assets and operations enables management to make faster and more precise decisions regarding business operations. Additionally, costly problems and problem processes can be identified faster. Ultimately leading to a more efficient and competitive organization.

RFID versus bar codes (with CMMS). Using a radio signal, rather than a light beam, RFID tags contain a globally unique identifier and memory, which can be read from long distances, at high speeds, and in virtually any orientation. This flexibility and its numerous advantages over bar codes, make it ideal in virtually any industry for tagging, tracking, and managing assets. See Fig. 2.40.

Figure 2.40 Example of a standard bar code.

Bar codes, unlike RFID, have many limitations which restrict their use and limit their effectiveness in many operational environments. Some of the more prominent limitations of bar-code technologies that RFID technologies overcome include:

- *Required line of sight.* Bar codes are scanned with a beam of light (using optics) and therefore require that the complete bar-code image be intact and fully visible is order for the bar-code information to be interpreted correctly.
- *Susceptibility to damage and dirt.* As bar codes rely on line-of-sight optics to interpret the product code, damage to the surface of the bar code or having the bar code covered in dirt can prohibit the reading of a bar code and therefore render the system useless in harsh and rugged environments.
- *Static information.* The information that is printed on a bar code cannot be modified. During its lifetime an asset or object may experience many status changes or require additional information to be associated with it. Using bar-code technology, this can only be accomplished if the software application has the functionality to allow the database record associated with the particular bar code to be modified. This can become particularly difficult for isolated or remote field activities.
- *Limited unique identification capabilities.* Unlike RFID, which can provide trillions of unique serial number combinations, most bar codes only allow for a few million unique serial identification numbers to be individually represented. That is why most bar-code systems group similar objects and assign a product code rather than identifying every object uniquely.
- *Lack of security.* Bar codes can very easily be copied (using a photocopier or digital camera), and then these secondary copies of the bar codes can be used to input data into a CMMS that is not accurate and can cause costly problems in the management of equipment and facilities. (e.g., A particular company was only getting 50 percent of the life cycle out of certain expensive pieces of equipment due to maintenance workers who were forging the maintenance records. They had photocopied the bar codes that were attached to the equipment and would stay in their truck checking off the maintenance WO using their mobile computer.)
- *Human interaction.* Bar codes in nearly every instance require the involvement of a person to conduct the actual scanning. RFID in many instances can be setup to automate the scanning process and eliminate the requirement for human intervention. However, the use of RFID technologies in conjunction with a CMMS, will still in most cases necessitate having a human in the data collection loop.

Bar codes do however have some advantages over RFID with respect to the current market offering.

- Labels are very cheap—just cents each.
- Business and use cases for bar codes are very well defined.
- Bar code technology is less complex than RFID technology.
- Great diversity of hardware offerings and supporting computing infrastructure.
- Software integration tools are much more mature, developed, and readily available in the market place.
- Many software applications, including CMMS, can be purchased with support for bar-code technologies as an inherent part of the offering.
- The amount and level of expertise available in the market place for bar-code technologies vastly surpasses that of RFID technologies.

Overall however, RFID is definitely a superior technology and will eventually replace bar codes in many application sectors. This is mainly due to many of the advantages it holds over bar codes, several of which are listed here:

- No line-of-sight required.
- Long read ranges allow for wireless tracking.
- RFID tags can be packaged for harsh and intrinsically safe environments.
- Data storage on the tag, allows the asset to "tell you about itself and its history."
- Tags are passively powered (do not require a battery), and thus have a practical infinite life span.
- Readers can capture RFID tag information without the intervention of a human user.

Future of RFID with CMMS. Currently there are very few instances of RFID technologies being used with CMMS. For the next couple of years most of the integrations of RFID technology into CMMS will result from individual end users who take the initiative to move toward advanced and innovative solutions for their maintenance management problems within their workplace.

Additionally, CMMS providers will begin limited offerings involving RFID to solve selected client-defined problems. It is predicted that RFID offerings as part of a CMMS offering will become increasingly common within the next three to five years. RFID will not begin to proliferate and become an industry standard in the CMMS application space until many of the major providers begin incorporating RFID into their product offerings.

Also, as the technology becomes the industry standard for automated data collection and successful use cases become more prevalent, equipment manufacturers will begin incorporating RFID tags and associated equipment information as part of their standard offering. This will mean that end users who purchase and use the equipment will not need to retrofit RFID tags on the equipment and it will lead to much closer relationships between equipment manufacturers and providers of asset/equipment management technologies.

The final large influencing factor that will lead to the widespread adoption of RFID technology in the CMMS field will be the continued growth and expansion of mobile computing and real-time connectivity. As computing devices that can accommodate RFID technology begin being used much more profusely so will RFID technology itself. Furthermore, the ability to link data to the centralized CMMS database in near real time will justify having a means of collecting data more rapidly and accurately.

Conclusion. As it has been illustrated in the aforementioned sections, RFID technology can be an immensely valuable addition to any CMMS. However, even though RFID has been around for many years its practical application use is still only relatively new. This technology has the potential to become one of the most prolific technologies in history especially with the recent mandates from Wal-Mart, the U.S. Department of Defense, and the FDA. There is no doubt that advancements in RFID technologies in the larger retail, defense, and pharmaceutical sectors will lead to the widespread adoption of this technology in nearly all sectors of business and will soon become the standard means of automated data collection.

With respect to RFID integration with a CMMS, there are a couple of key considerations that any company or end user must understand.

First and foremost is the fact that it is essential when introducing a new technology into an existing process that the process must also adapt and change in order to reap the benefits of the new technology. The same holds true for RFID and CMMS. This is an extremely important point that must not be overlooked. As part of the planning process for incorporating RFID technology into a CMMS, the process for using the "new" system must be mapped in detail and completely understood by all stakeholders. Otherwise the technology will probably not be used properly or effectively, you will not receive buy-in for the new system from the field workers and the integration could become costly and experience time and budget overruns.

Second, unless your company has engineering staff and software developers, who are very experienced with RFID technologies, the use of an experienced RFID solutions provider or systems integrator is strongly recommended. An RFID solutions provider can be extremely important in ensuring the success of your RFID system deployment. An RFID solutions provider can assist or lead the RFID integration in nearly every aspect including:

- Selection of the most suitable RFID technologies for your short- and long-term requirements
- Installation of readers
- RFID tag mounting and placement
- RFID tag registration and enrollment
- Provision of RFID middleware and integration tools
- Process development for system use
- System training and best user practices
- Managing expectations within your organization

The companies which will benefit most from RFID technology will be the early adopters gaining market advantages via process improvements, increased time and cost efficiencies, and innovative product offerings that create solutions for industry-wide problems, which in-turn will lead to increased sales and revenue growth.

Speech-enabled maintenance management*

Introduction. Significant advances in speech-based technologies have emerged to provide computers with the capability to cost-effectively recognize and synthesize speech. Additionally, wireless communications have ascended to where the number of mobile phones will eclipse land-based phones and the Internet has become a commonplace communication mechanism for businesses. The confluence of these technologies portends interesting opportunities for maintenance management.

Maintenance, by its very nature, is a highly mobile activity. This mobility requirement constrains a craftsman's ability to receive and provide information that can improve productivity, reduce costs, and improve overall management of the maintenance process. Once the workers venture beyond their wired environment, their options to gain access to information resources diminish. Figure 2.41 provides a lighthearted perspective on the challenges of remote access for the mobile worker.

An additional factor is that maintenance workers, along with other "skilled" practitioners such as physicians, often view computer technology

*This section is contributed by Michael Dougherty, Crossbar Solutions, LLC, mdougherty@crossbarsolutions.com.

Figure 2.41 Remote access.

as extraneous to the job at hand. Their affinity toward, as well as exposure to computer applications is often less than optimal for maintaining and fully utilizing information resources.

Organizations realize that management of the maintenance process can lead to significant cost savings as well as productivity improvements. Trends toward planned and scheduled maintenance programs to improve efficiency and effectiveness have prompted maintenance organizations to deploy CMMS. While a CMMS is the backbone for automating the management approach, efficiently capturing data and providing easy access for system users are key success factors for the system.

Technology, standards, and specifications. There are many technical components to speech-enabled applications. A key point to be emphasized here is that the applications and technology discussed fall into the category of *speaker-independent*. What this means is that the system does not need to be trained, it is independent of the speaker. This differs from the some of the more familiar "dictation" systems such as IBM's Via Voice and ScanSofts's Dragon Naturally Speaking. With speaker-independent systems you do not need to train the computer application by speaking words or phrases. This reduces the cost of deployment and increases user acceptance. Some of the major speaker-independent software vendors in the market include IBM, Scan Soft, Motorola, and Microsoft. Scan Soft has recently acquired two of the leading speech-technology vendors: Speech Works and Nuance. Microsoft's support for the tablet-PC along with its push into the speech-domain has added some additional strength and alternatives to deploying speech-enabled applications. Information on the ROI advantages related to speech-based applications can be downloaded from their Web sites. Another important source of information can be found at http://www.voicexml.org.

There are a number of technical details related to fully developing and deploying a speech-based application, however one of the primary technological components, VoiceXML, bears mentioning. Simply put—VoiceXML is an open standard used by developers to build speech-enabled applications. It has evolved out of the same technology architecture that produced HTML, which led to the rapid adoption of the World Wide Web over the Internet.

VoiceXML version 2.0 was released in 2004 as a W3C (World Wide Web consortium) standard. Many of the voice recognition software companies support version 2.0, and planning for version 3.0 is in the early stages. The importance of this technology is that it will allow companies to build platform-independent speech-enabled applications and leverage a company's investments in web-based applications.

While not a W3C standard, the *speech and application language tags* (SALT) specification is also a viable consideration for the underlying speech-recognition platform. This specification was put forth by Microsoft, Intel, and several other companies. Some industry experts foresee a time when VoiceXML and SALT will combine into a single standard, especially in the area of multimodal applications.

Deborah Dahl's article, "Guide to Speech Standards," in the March/April 2005 issue of *Speech Technology* (www.speechtechmag.com) provides a comprehensive summary of current speech technology and standards. Figure 2.42 shows the major elements needed for speech-enabled applications.

The other technical components or subsystems that round out the requirements include a network interface; a telephony (i.e., telephone) interface; a *text-to-speech* (TTS) engine that translates computer text into spoken words; a speech-recognition engine that translates spoken words into computer text; and an audio subsystem to record and play back audio files. Because radio communications are prevalent in many CMMS environments, it is worth noting that recent advances in the industry have integrated VoiceXML capabilities with two-way radio infrastructure. This allows companies to leverage their current two-way radio

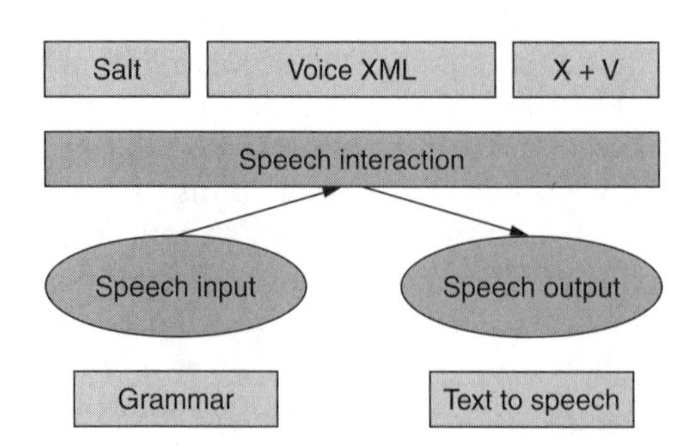

Figure 2.42 Major components needed for speech-enabled CMMS.

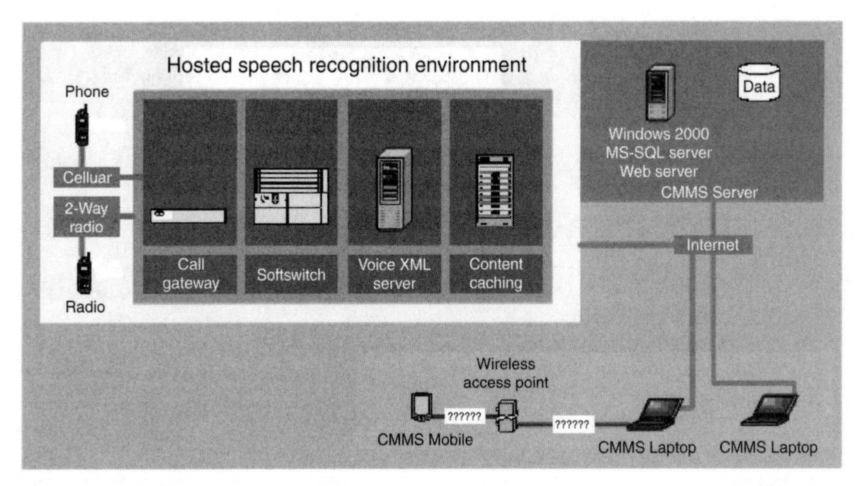

Figure 2.43 Technology components.

equipment, and supports legacy trunked and conventional base stations and two-way repeaters. Figure 2.43 provides an overview of the major technical infrastructure components.

Speech-enabled CMMS. A compelling feature of speech-enabled maintenance management is that users can interact with applications to retrieve information and keep the systems up-to-date via standard or mobile phones. A prototypical dialogue between a maintenance worker (M) and a speech-enabled application (C) would sound something like shown in Fig. 2.44.

In this scenario, Marshall is a maintenance foreman on his way back from inspecting a failed piece of equipment. Rather than wait until he returns from the site, he is able to call in his information to the application and approve the WO for dispatch to a maintenance crew.

A few years ago, a scenario like this would have been highly improbable, but the technology exists today to build solid, fluent speech-enabled applications. The features available in speech technology can adequately deal with the complexity of human speech. For example when the foreman responds, "one zero zero one" (1001) to the WO number question, the speech-recognition engine could have also recognized "one thousand and one" or "ten o one."

Another example of the robustness of current speech technology is in a feature known as *mixed-initiative dialogs*. This allows a user to make a statement such as "set the failure class to pumps." The application can strip out the two key pieces of information "failure class" and "pumps." This is analogous to filling out a form on a computer screen.

Voice	Dialog
C	Welcome back Marshall. Your choices are work orders, equipment, inventory, or more choices
M	Work orders
C	Update, create, work order details?
M	Update
C	What is the work order number?
M	One zero zero one
C	What would you like to update. Say "Choices" to hear a list of options.
M	Set failure class to pumps Set problem to low volume Set status to approved
	Silence
C	Are you done with this work order?
M	Yes
C	Anything else?
M	No
C	Goodbye.

Figure 2.44 Dialogue between a technician and CMMS.

Mark Agan, a Director with Professional Services Facility Management of Fort Washington, Pennsylvania, provides facility management and maintenance services to hospitals and universities. They currently use a standard voice response application to help dispatch services. "Right now the callers have to punch in their location code and the required service using the phone keys. We then dispatch a maintenance worker to the location via a beeper. However, with speech recognition, the staff could say 'water leak in room 407.' The advantages are that the caller can interact in a more natural manner, we can respond to many more types of requests, and we can also load the request directly into our work order database. This would let us do a better job managing work requests."

Elements of a speech-enabled application. Users of current MS-Windows or web-based applications will quickly recognize that speech-enabled applications are in many ways similar to traditional computer applications. The main components of speech-enabled application include dialogs, navigation, forms/fields, and grammars. These compare with the modules, menus, forms/pages, and valid value lists of traditional applications.

Dialogs. These are the main components of a speech application. They group together the similar conversational elements of an application much the way functions are grouped into modules within a traditional computer application. An example of this would be a WO module versus a storeroom module in a CMMS application.

Navigation. Just as users can click on menu items within a computer application, speech-based applications provide a similar audio menu to the caller. The user is then transitioned to another submenu or a specific form. An important feature in speech applications is the ability to provide links directly to any portion of the application without traversing the audio menus. This feature, known as "barge in" allows the caller to jump directly to the desired function. This allows experienced users to use the system more efficiently.

Forms/fields. Speech-based forms serve the same purpose as traditional computer application forms. They provide the user with fields in which data can be entered or retrieved. As mentioned earlier, mixed initiative dialogs allow the caller to fill out the form more naturally.

Grammars. The primary role of grammars is to predefine what a user can say and that the application can recognize. This is similar to the valid value lists that users pick from when entering data into an application. It should be noted that grammars have a broader context in speech applications, as the user must also specify field names along with their values. In the earlier example where the speaker said, "Set the failure class to pumps," the term "failure class" maps to the field name and the term "pumps" maps to the field value.

Considerations and limitations. While there are obvious advantages to *voice user interfaces* (VUI, pronounced vooee), they do have limitations when compared to the more traditional *graphical user interfaces* (GUI). Some of the limitations of speech-based applications include:

- People can read much faster than they can speak. For example, listening to a long list of open WOs over the phone is not practical.
- People have a difficult time remembering what they just heard. Consider what happens when you stop to ask someone for driving directions. Speech applications are not efficient if unfamiliar data needs to be committed to short-term memory or repeated continuously.
- People can say anything, but computers are limited in scope. Because a conversation between two individuals is unbounded, people will naturally infer that a conversation with a speech-based computer application is open-ended. However, the speech-based application must be carefully constructed to guide a user through the valid options.

Other considerations for the maintenance management world are the noisy environments where maintenance is often performed and the prevalence of two-way radio communications that limit some of the functionality of speech-based applications. Several years ago, Professional Services tried to deploy speech recognition to help inventory the equipment at customer locations. "Some of the areas like boiler rooms were just too noisy," said Mark Agan. However, as the technology improves to filter out background noise, properly designed speech applications will expand their reach.

And while talking over the phone is a task anyone can master, talking to a computer application still presents challenges. Professional Services still sees some of the functions being handled by more sophisticated users. "On some of the data entry activities, you still need to have knowledge of what the application is supposed to do. You simply cannot hide all of that from the caller or explain it all over the phone."

Multimodal applications. Another important development in the field of speech-enabled applications is in an area known as multimodal applications. While traditional telephones provide a single mode of communication, that is, voice, PDAs and smart phones can provide both voice and data at the same time. In the CMMS world, multimodal allows the user to say "list supplies" and then view the 5 to 10 items that are needed for the job on their PDA versus listening to the list over the phone. The multimodal approach eliminates some of the significant user interface challenges when working with PDA-type devices.

Anyone who has used a PDA understands that working with the small keypads, screen sizes, and stylus pens can be tedious. Consider the following example when a user is trying to set the dates for a WO:

Multimodal Approach. See Fig. 2.45.

PDA Approach. See Fig. 2.46.

Multimodal, or text and data, is considered by some as the panacea of speech-enabled applications targeted at workforce automation. The major challenge with multimodal is how much speech recognition can and should be done on the PDA device. As the processing power of the PDA improves, this will become less of an issue, but for now it remains an important design consideration and can limit the robustness of the application. Some vendors have taken the approach where all of the

Voice	Dialog
C	Marshall, what is the expected strat date and duration for this activity?
M	Start next Wednesday, duration 1 week.
C	You said the activity would start on November 2nd and would last one week. Is this correct?
M	Yes

Figure 2.45 Multimodal approach.

PDA	Dialog
C	Position cursor in expected start date field
M	Tap the calendar icon Tap next month (since it's October 31st) Tap November 2 Tap OK (to accept date) Tap the duration field Tap the digit icon to show a grid of numbers Tap 1 on the grid Tap the duration units field Tap week from the selection list

Figure 2.46 PDA approach.

speech recognition is done on a remote server. This works well if the user can establish a wireless connection with their PDA. As wireless network connectivity grows this approach could offer the most cost effective solution, since the application will be on the network and the PDA will serve exclusively as the interface.

Benefits of speech-enabled CMMS. From a CMMS perspective, the key questions on most managers' mind would be: What benefits will I get from speech-enabling application? How much will it cost? And how do I get started?

Before you think about the cost of a speech-enabled CMMS, there are a few prerequisites that you should consider. Some of these include:

- Do you have a stable CMMS in place?
- Do you have more than 10 maintenance workers?
- Are your maintenance functions distributed over a wide geographical area?
- Have you taken advantage of more traditional performance improvement approaches such as WO planning?

The benefits and costs of implementing speech recognition depend on multiple factors. The most pivotal is *scope*. A primary reason new technology initiatives fail is that they are too big, too complex, and take too long to deliver results. Keeping your initial project to a manageable size will increase your chances of success and minimize your investment risk. Speech can be applied to a variety of CMMS areas so you should break CMMS into functional categories and potential speech-enabled applications. Some of these are as shown in Fig. 2.47.

Figure 2.48 details the major functional areas within a CMMS application from a graphical perspective. Each of these areas provides potential opportunity for improving performance with speech.

Voice	Dialog
Dispatch	– Notify workers of their next assignment. – Confirm site visits with customers before your workers arrive on-site.
Work orders	– Let customers initiate a work request from their phone. – Update work order status, problem codes, or time estimates from the field.
Equipment	– Check on the status of a piece of equipment before traveling to a site. – Find out what spare parts, supplies, or sub assemblies are needed to work on a piece of equipment.
Tools	– Find out who has a specialty tool that is shared among crews. – Check out/check in tools from/to remote storage sheds.
Supplies	– Request items from the storeroom so that they are ready for pick-up when you arrive. – Check on item availability so you don't waste time traveling to the supply room only to find out that the item is not in-stock.
Crew	– Enter time and attendance data from the field. – Find out who has a particular skill set and if they are available.
Safety	– Find out if there are safety hazards associated with the task or the equipment. – Verify the safety supplies needed for a task.

Figure 2.47 CMMS functional categories.

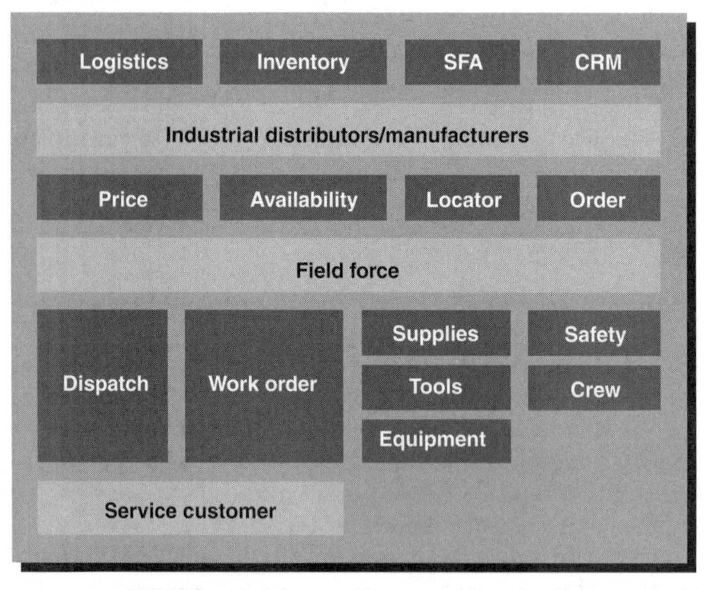

Figure 2.48 CMMS functional areas. (*Courtesy of Crossbar Solutions, LLC*)

When considering speech, look for hard and soft benefits. This process will not differ much from the process used to justify a CMMS. Some of benefits could include:

- Increase resource productivity from 5 to 25 percent
- Achieve better allocation of scarce resources
- Improve profitability by avoiding costly downtime or unavailability of resources
- Provide superior customer service

As an example, suppose that you provide maintenance services to a large hospital facility and you need to provide a more efficient mechanism for hospital staff to open work requests. Using a speech-enabled work request system you could eliminate the data entry costs of keying several thousand work requests into your CMMS. This can be handled with a modest investment and could provide substantial benefits via improved response times and customer service. A small application such as this could cost between $15,000 to $25,000 to design, develop, and implement.

A leading cost consideration is whether you bring the entire technology infrastructure in-house or you use a hosted solution. Opting for a hosted solution can save $20,000 to $50,000 on the initial investment. One example of a company that provides a hosted VoiceXML environment is Voxeo out of Orlando, Florida. They provide all of the telephony, speech-recognition, and VoiceXML infrastructure; and all you need to provide is an Internet connection to your application server. You pay a monthly subscription fee, which is comparable to standard cell phone costs.

A word of caution, a significant area of investment could be related to the state of your existing data. Suppose that you have coded all of your locations with six character codes. So instead of BOLIER ROOM you have BOLRRM. You will need to map the existing codes to a language definition. This will add time and cost to deploying speech. Depending on your scope, you will need to make sure that this is fully sorted out in the requirements phase.

Getting started with speech-enabled applications is no different than getting started with any IT project. The key steps include:

- Identify your needs/high-level requirements
- Develop a business case
- Select an area of high benefit and reasonable cost
- Develop the detailed requirements and design specifications with the end-users

- Review cost/benefits
- Construct the solution and integrate with your CMMS
- Implement a pilot and measure performance
- Implement the whole project

Conclusion. Major organizations in the airline, financial service, and transportation markets have successfully deployed speech-based applications to provide higher levels of customer service as well as save millions of dollars in operating costs. Some of the industry segments that could benefit from speech-enabled maintenance management include utilities, municipal transportation authorities, academic institutions, medical facilities, and retail chains.

As organizations look to make their workforces more efficient by implementing a CMMS, they need to address the accessibility of these applications by a mobile workforce. While arming them with laptop computers, PDAs, and other hand-held devices can surely help; the low-cost, ease of use, and ubiquity of voice communications presents an alternative that is hard to ignore.

Organizations that want to ensure that their investments in CMMS applications are fully utilized should begin looking at how speech-enabled maintenance management can extend current applications to workers who are highly mobile and need to stay that way.

Appendix 2A Machine Replacement Analysis*

A problem frequently encountered in maintenance management is the optimal time for replacing a machine. An existing machine or equipment needs replacement for a variety of reasons such as deterioration, costs, obsolescence, excessive downtime, new requirements of the marketplace, and risk of catastrophic failure.

As a machine ages, the cost of maintaining and operating the machine tends to increase, resulting in less efficient operations. Routine maintenance, energy, and labor costs required to operate the machine goes up with continued use over a long period of time. Oftentimes, the adjustments on older machines are difficult to make resulting in inconsistent and lower quality products with significant market implications. Newer machines tend to have lower operating costs and consistent quality levels. Advances

*This appendix is contributed by Bharatendu Srivastava, Marquette University, bharat.srivastava@marquette.edu.

in technology and automation can also result in superior machines with lower production costs resulting in frequent equipment replacement. A consequence of recent globalization of markets and rapid product innovation is the ever-shrinking product life cycles resulting in frequent changes for products and services in the marketplace. Thus, a replacement decision may stem from one or several of the mentioned factors.

Replacement analysis as discussed here is based on the concept of economic life. The basis for this analysis is that when a machine is put in service, it has associated costs such as depreciation, operational, and maintenance. In general, the cost of depreciation decreases with each year of operation whereas the operational and maintenance costs increase with age. Economic life is the life at which the average annual cost of operation is at its minimum, which means a machine should be replaced when it reaches its economic life. This outcome is valid only in an ideal situation, for it is based on the assumption that the new machine is identical to and cost the same as the existing machine. In reality, this may not be true and it is possible that it may be cheaper to operate the existing machine beyond its economic life than purchasing a new machine.

Replacement analysis is a choice between mutually exclusive alternatives. Either the existing equipment will continue to be used, or new equipment will be purchased to replace it. Formally, for our discussion, the economic life of a machine is defined as the number of years at which the *equivalent uniform annual cost* (EUAC) is lowest. Generally speaking, economic life of a machine is often shorter than either the physical or useful life. In analysis that follows, we make use of the following concepts.

1. Market value of a machine is the actual price or value of the machine. Obtaining an estimate from the marketplace as determined by the sale of an asset (or comparable asset), is generally the preferred method. Estimates obtained from expert appraisers, engineers, or accountants are also acceptable. Often, manufacturers and trade associations also maintain data on the market values of industrial machinery.
2. Because costs accrue at different points in time during the life of a machine, we utilize the following formulae from the area of financial mathematics.
   a. Single payment present worth factor

$$(P/F, i\%, n) = P = F[1/(1+i)^n]$$

   b. Capital recovery factor

$$(A/P, i\%, n) = A = P\left[\frac{i(1+i)^n}{(1+i)^n - 1}\right]$$

Note:
In the above notation $(X/Y, i\%, n)$: X = what is sought, Y = what is given, i = interest rate, n = number of periods.

P = present sum of money

F = future sum of money

A = end-of-period cash flows in a uniform series continuing for n periods, the entire series equivalent to P at interest rate i.

In all analysis, end-of-year convention is followed.

3. Interest rate

Economic life of a new machine

The marginal cost (MC_k) of extending the service by an additional year depends on the market value of the machine and the annual operational and maintenance expense. In general MC_k for year k is defined as:

$$MC_k(i\%) = (1+i)\ \text{market value}_{k-1} - \text{market value}_k + \text{annual operating and maintenance expense}_k$$

Marginal cost consists of loss in market value, foregone interest from selling the machine, and the annual operational and maintenance cost.

Once MC_k has been computed, $EUAC_k$ is:

$$EUAC_k = \left[\sum_{j=1}^{k} MC_j(P/F, i\%, j)\right](A/P, i\%, k)$$

Example Initial investment for a machine is \$12,000 and its expected market value, annual operational, and maintenance expenses in each year are as given. Interest rate is 10 percent per year. What is the economic life of the machine? See Table 2A.1.

In the Table 2A.1, $EUAC_2$ and $EUAC_3$ are calculated as:

$$MC_1 = (1.1) \times 12{,}000 - 8000 + 500 = \$5700$$

$$MC_2 = (1.1) \times 8000 - 6000 + 800 = \$3600$$

$$EUAC_1 = MC_1 \times (P/F, 10\%, 1) \times (A/P, 10\%, 1) = 5700 \times 0.9091 \times 1.1000 = \$5700$$

$$EUAC_2 = [MC_1 \times (P/F, 10\%, 1) + MC_2 \times (P/F, 10\%, 2)] \times (A/P, 10\%, 2) = [5700 \times 0.9091 + 3600 \times 0.8264] \times 0.5762 = \$4700$$

TABLE 2A.1 Initial Investment Example 1

End of Year	Market Value ($)	Annual Operational and Maintenance Expense ($)	Marginal Cost ($)	EUAC ($)
0	12,000	—	—	—
1	8000	500	5700	5700
2	6000	800	3600	4700
3	5500	1000	2100	3914
4	4000	2000	4050	3944
5	3000	5000	6400	4346

Based on these calculations, the machine will have a minimum EUAC of $3,914/year if it is kept in service for three years. Therefore, the economic life of the machine is three years.

Existing machine's economic life

An analysis similar to one done for a new machine can also be used to compute the economic life of an existing machine. The decision rule often used in such situations is to keep the machine as long as the marginal cost (MC) of operating the existing machine is less than the best EUAC for the new machine. The following example illustrates the procedure.

Example Suppose the machine discussed in the previous example is now two years old and there is no change in the market value and the annual operational and maintenance expense of the machine. Its marginal cost (MC) remains as computed before. This machine can however, be replaced by a new machine for which the data is given in Table 2A.2. The new machine costs $15000 and its expected market value and annual operational and maintenance expenditure is shown in Table 2A.2. Interest rate is 10 percent.

Both MC and EUAC are computed in exactly the same manner as before.

The minimum EUAC of the new machine is $4427/year for four years of service.

Given that the existing machine has already been in service for two years, its marginal cost for years three to five of operation are $MC_3 =$ $2100, $MC_4 =$ $4050, and $MC_5 =$ $6400.

Because MC_3 and MC_4 are less than the minimum EUAC = $4427, the existing machine should be used for another two years and then replaced with the new machine because at that point in time $MC_5 =$ $6400, is greater than the minimum EUAC.

TABLE 2A.2 Initial Investment Example 2

End of Year	Market Value ($)	Annual Operational and Maintenance Expense ($)	Marginal Cost ($)	EUAC ($)
0	$15,000	—	—	—
1	10,000	$550	$7050	$7050
2	8000	850	3850	5526
3	7000	1200	3000	4763
4	6500	2000	3200	4427
5	5000	4000	6150	4709
6	4500	6000	7000	5005

Our replacement analysis has been limited to before-tax consideration. After-tax consideration complicates the analysis and depends upon the depreciation method, capital loss or gain, tax rate, and company policies. However, it is important to mention that the methodology is similar.

Appendix 2B ISO/QS Compliance Case Study*

Having ISO/QS certification gives you the right to *lose* it every 6 months! This is a rather drastic evaluation, but quite frankly, a realistic one. Maintaining the rigid quality system standards established by the international committee dictates periodic reevaluation of a company's adherence to those standards. Simply stated, the philosophy behind ISO is:

Say what you do	Document *standard operating procedures* (SOPs)
Do what you say	Adhere to these SOPs
Prove it	Document the two previous statements!

Those of us in the manufacturing sector have become familiar with the requirements of managing and maintaining our portion of the ISO/QS certification for our business. ISO standards management for maintenance-related functions can be accomplished in your *computerized maintenance management system* (CMMS) with very little additional work on your part.

WCI Steel began computerized management of its assets, maintenance responsibilities, inventory, labor force, and preventive maintenance (PM) in 1990, selecting MAPCON over a host of other products. This report examines the solutions their CMMS provided in maintaining an ISO/QS-9000/2 certification.

*This case study is contributed by Edward T. Johnson, WCI Steel, etjohnson@wcisteel.com.

Say what you do

An existing table in the software provided an excellent vehicle for creating and editing the control processes for our maintenance environment. SOPs for both maintenance and operations can be developed. The W Y S I W Y G format was compatible with "cut and paste" features of Windows. We are also able to attach detailed schematics, diagrams, and even photos to aid in the safe completion of a repair. Supplementing this is an import/export feature that can be employed with existing files in other systems. Either way, you gain several significant advantages by managing these records in a plant-wide CMMS, as opposed to other established processes:

Plant-wide, on-screen access to procedures. Many companies manage their SOPs via the department secretaries/clerks who keep the records on their PC or saved to a network drive. The files are restricted to one or two employees for security purposes. Copies of the most current revision are made and distributed throughout the plant or department, and placed in "books" for ready access and review. Since out-dated information is a definite noncompliance in ISO/QS management, procedures have to be developed to define the handling of the old copies removed from each book. Folders, files, storage cabinets—all these steps require dedicated, labor-intensive management to insure conformity.

A multileveled security system in a CMMS provides the same control features to create/edit records, plus allows on-screen viewing of the most up-to-date changes to procedures anywhere in the plant. All this happens within seconds after the changes are made! No more need to make, distribute, retrieve, and manage copies of your procedures.

By managing SOPs in a CMMS, there are no out-dated, obsolete issues to address; the most current revision is the one displayed. Browse screens, or screen-editing security guarantee the integrity of the data.

Not all procedures are ISO procedures. The ability to customize screens and menus in a CMMS is a valuable tool in the management of ISO/QS procedures. By creating a new dictionary item in the file, it is possible to add a Boolean-formatted field, and place it on the screen as a "required" entry. You are now able to list all ISO and nonISO-related procedures via ad-hoc reporting features. Build yourself a new lookup, and you can sort the records according to their Y/N ISO "flag." Hands-on users can quickly find appropriate procedures when needed for review, or as part of an audit.

Job safety analysis (JSA) management. In most manufacturing environments, *safety procedures* are defined for virtually every maintenance function performed. Documentation of these procedures provides valuable insurance that the job gets done safely and correctly. In many cases, these procedures are a matter of law, and routine discussion with employees is part of an effective safety program.

Whether they are called JSAs or safe maintenance procedures or job tasks, they all basically outline the step-by-step procedures required to safely and correctly do the job. These procedures are the basis for many ISO/QS procedures for maintaining equipment.

As with corporate or departmental ISO procedures mentioned earlier, JSAs are commonly managed in one location by one or two employees for security purposes.

The same benefits can be reaped from a full-functioned CMMS as well; W Y S I W Y G format, plant-wide access via PCs connected to your network, create/edit security, current information. Safety procedures can be attached to equipment records or to PM routines as well. When work orders for the equipment are generated, you have the ability to print a copy of the safety procedure(s) as part of the job.

Identification of ISO-critical euipment. A basic requirement of ISO/QS certification is the identification of those pieces of equipment in your facility that have "a direct impact on the quality of the product." If employees do not know which pieces of equipment impact the quality of your product, it is doubtful that you will be able to maintain it to provide consistent quality. Once again, computerized maintenance software provides a solution.

As mentioned earlier, one of the strengths of some CMMS programs is giving the user the ability to customize the program to meet their needs. Coupled with a responsive support service from the developer, user-defined fields can be added to data entry screens. A required Y/N entry enables a company to answer the question "What ISO-critical equipment am I responsible for?" Figure 2B.1 demonstrates this problem-solving feature.

"Qualified" maintenance. Once an organization identifies which equipment is ISO-critical, the next phase of the process is developed. *Who* is responsible for maintaining the equipment, and the *skills required* to do so need to be addressed. Human resources, craft codes, and other integrated modules provide the solution.

Craft codes are defined in the program, and employee records are entered in a human resources file. By defining specific training programs in an appropriate table, programs can be attached to individual employee records—either through timecard screens or directly from the human resources table. Entries made via the timecard system should automatically update the H/R record to show attendance at training

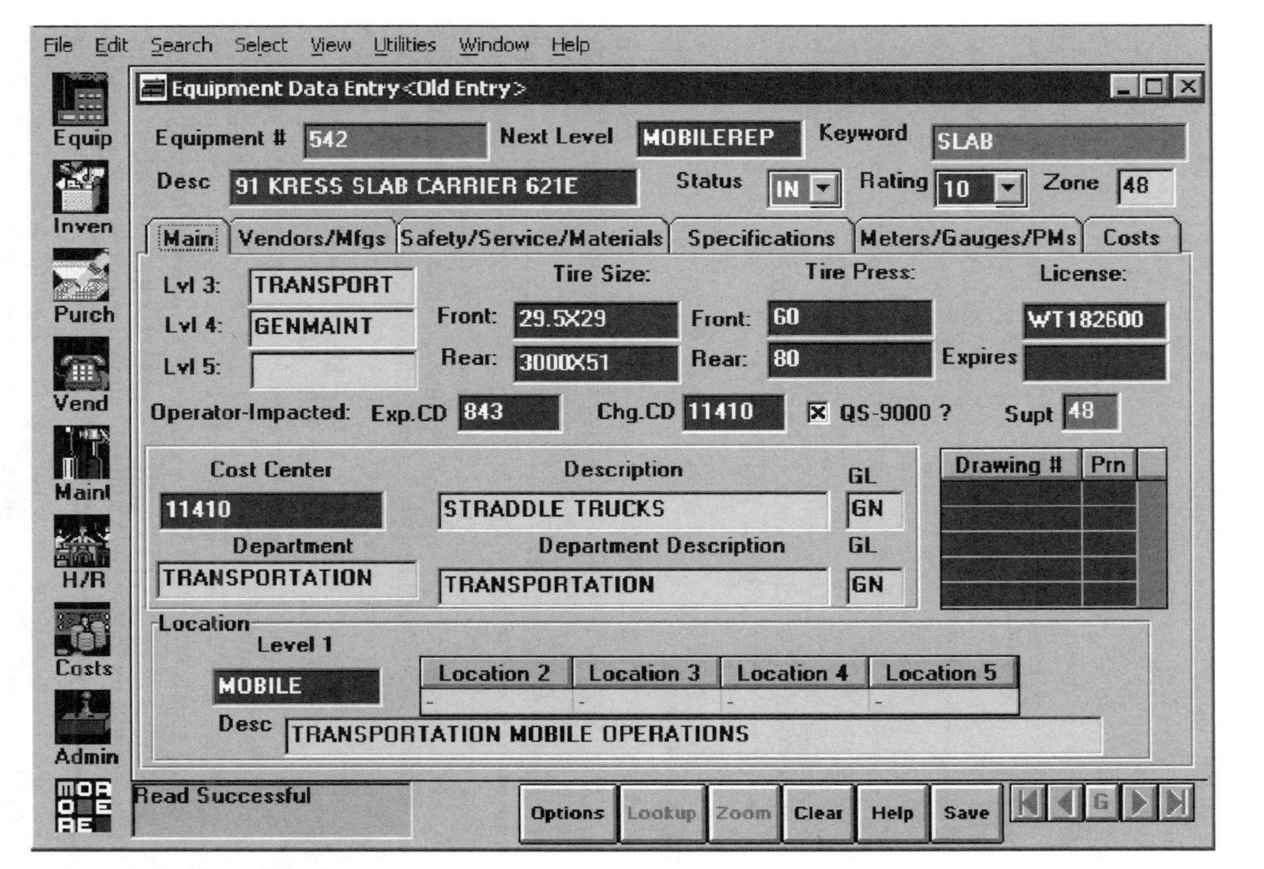

Figure 2B.1 Problem-solving feature.

sessions. Again, the reporting features throughout your software provide you with the tools to respond favorably to an audit.

Managing critical spares. It is not enough to know which equipment is critical, who will maintain it, and what skills are required. Reality tells us that nothing lasts forever, and parts eventually wear out. Keeping the equipment on line means having critical spare parts available in the event a breakdown occurs. When this happens, you must ensure that "shortcuts" are not taken that will compromise your ability to keep the unit performing efficiently to maintain product quality.

Inventory tables, coupled with a vendor database, provide maintenance organizations with the tools to ensure they have the right part when needed. Part status ratings called CR (*critical*) or CS (*critical spare*) enable maintenance groups to manage items in more than one location in the plant. Or, you can create a single stockroom called "critical_spares" and group all essential or long lead-time parts there. Reporting could then reflect your ability to make essential repairs to critical equipment in a timely fashion.

Do what you say

Now that you have identified the *what*, your focus will be turned to *who* and *how* (say what you do), and finally the *when* aspect of ISO/QS management, or the "do what you say" step.

PM is the minimum amount of resources required to maintain optimum equipment performance. Changing the oil and filter in the family SUV every week—although certainly a benefit for the engine—is hardly PM. The same applies to your gear reducers, motors, cranes, conveyors, and process lines. Every hour spent maintaining equipment affects profitability. On the other hand, spending too few hours on maintenance affects your equipment negatively, and seriously impacts its ability to maintain quality. That is where PM pays off.

Once the *what, who*, and *how* have been determined, the *when* becomes the job of the PM module. PM procedures detail the minimum steps required by the equipment manufacturer to maintain equipment uptime. The automotive industry's three months or 3000-mi oil change interval is based on "normal" operation. Likewise, manufacturer's minimum standards for your equipment should be adjusted accordingly.

QUESTION: When does a company consider its PM system to be finished?
ANSWER: Never! Remember the saying - If you keep on doing what you're doing, you'll keep on getting what you're getting.

Once PM procedures have been developed and tested, it is time to turn the process over to the PC. Having established the "date last done" for all your equipment PMs, the computer assumes the responsibility of keeping track of the next time each maintenance function—or PM—is due. It does this by issuing a work order created from the information on the PM. Short-circuit this step, and you will certainly jeopardize your certification, and lose the "do what you say" piece of the process!

Work orders are generated from the PM procedures you have written, and assigned to the appropriate craft personnel (or defined crews) for execution. Once the work has been completed, the work orders are closed, with supporting comments and details noted. As an organization's work orders migrate from "reactive" to "proactive," their equipment uptime increases, and costs ultimately decrease.

As mentioned earlier in this report, JSAs or safety procedures can be attached to a PM that further detail the maintenance work required to safely keep the equipment running. Having a system that allows the user to draw information from several sources—without the need to duplicate in each—is certainly an asset provided by some CMMS packages. This report started out discussing the ability to create/manage SOPs in available tables. These records can be easily attached to a PM procedure without retyping the information! Further, if changes are made to the parent document, they will be reflected automatically the next time a PM they are attached to is generated.

Prove it! Developing and managing all of this information takes the combined efforts of several groups within an organization. First, a system manager oversees the operation of the program and is responsible for the training, implementation, and maintenance of the software. Next, departmental staff must build the human resource information and related training programs. The maintenance group must identify and develop the equipment and inventory information essential for successful management. Finally, a well-informed maintenance crew whose job is to keep the equipment running is a must. Each step of these processes begins to create the documentation that satisfies the "prove it" phase of ISO/QS-9000. Reports can be generated that answer the questions needed to verify compliance.

Because ISO/QS compliance involves equipment, inventory, training, maintenance and PM records, we decided to create a single menu where anyone involved in an audit could obtain information that demonstrates compliance. It is not uncommon for an auditor to ask for evidence of equipment identification, critical inventory, employee training, or PM compliance records from the person being interviewed. At WCI Steel, we have created a single menu in our software devoted to the "prove it" phase of ISO/QS management. Anyone involved in the audit process knows how to access this menu. It contains reports or "evidence" of compliance to the "do what you say" phase. The ISO/QS-9000 reports menu in use at WCI Steel is shown in Fig. 2B.2.

From this single menu, the department or individual being interviewed can answer questions like:

- What equipment is defined as ISO-critical in this department?
- Can you show me how your crew is doing on their PM requirements?
- How may times, and when, have you completed PM work on equipment No. ( )?
- What are your CS?
- Which employees have been trained in training program ( )?
- What quality-related SOPs exist in this department?
- Can you show me a copy of all the PMs for equipment No. ( )?
- Do you have any open PM work orders for the Millwrights?

Inquiries such as those listed involve equipment, inventory, human resources, work orders, and PMs. By consolidating all the reports to one menu, users do not have to know the layout of every report menu in the software, or determine if certain information is available from one section of the program instead of another. Our software allows us to create not only the reports that we need, but our own menus to display them from as well.

Appendix 2C Spare Parts Inventory Management*

An effective spare parts inventory management system is an important element of a good maintenance system. Poor availability of parts can lead to extended repair and downtime. On the other hand, optimal stocking of spare parts is a fairly difficult problem to solve. While forecasting parts requirements for a planned maintenance is relatively simple, it is the unplanned maintenance and breakdown of the machines that makes this problem so difficult. As such spare parts often have a low and erratic demand pattern with extremely high shortage cost. In some cases, inventory turnover for spare parts can be less than one turn a year. With rapid advancement in technology and innovation, the number of spare parts has also grown exponentially over the last decade. All of this makes forecasting, stock holding, and ordering policies for

*This appendix is contributed by Bharatendu Srivastava, Marquette University, bharat.srivastava@marquette.edu.

File Edit Search Select View Utilities Window Help
Equip
Inven
Purch
Vend
Maint
H/R
Costs
Admin
ISO / QS-9000 Reports Menu
Previous Menu
List ISO/QS-9000 Quality-related Eqp.
Critical Spares by Stockroom
List OPEN PM WO's - by Eqp. #
List Training by Employee
List CLOSED PM WO's - by Eqp. #
List Training by Craft Code
List Craft Compliance to PM Program
List All SOP's by Dept.
List ISO/QS-9000 PM's by Craft
List ISO/SOP's by Dept.
List PM Activity - by Eqp./Date Range
Browse PM Descriptions
List PM Activity - by PM#/Date Range
Browse PM Frequencies
PM File Report: Date Last Done
Print PM Description Forms
PM's Gen'd vs Comp. on Time - by Craft
List ISO WO's done - by Date Range
List Current PM Records by Craft
List ISO PM Frequency
Critical Spares by Department
Additional Caster ISO Reports
Critical Spares by Keyword
Options
Lookup
Zoom
Close
Help
Save

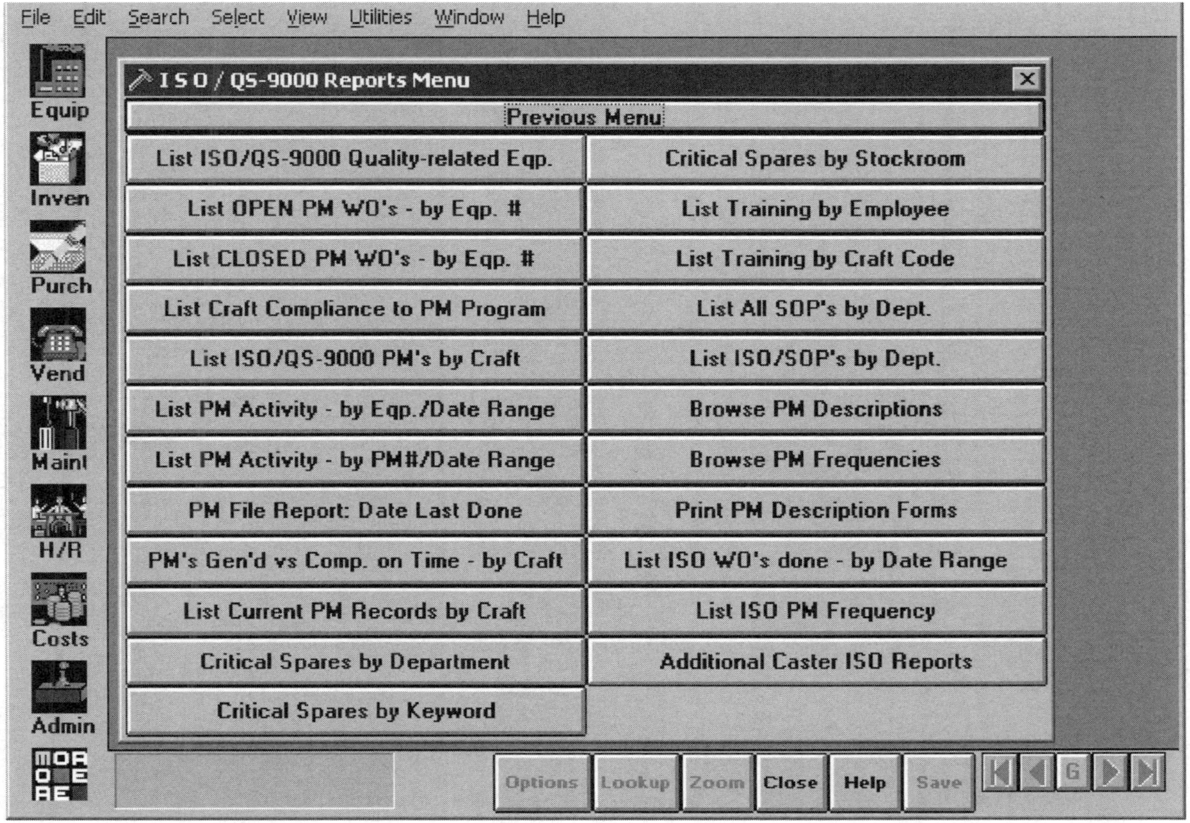

Figure 2B.2 ISO/QS-9000 reports menu.

spare parts a complex managerial undertaking. It is thus not unusual to encounter situations where there are either excessive amounts of inventory in the system or there are frequent stock outs. Both situations are undesirable as they may lead to excessive costs and poor service.

Some key points are described, which managers can utilize in dealing with issues related to improving the planning and control of spare parts inventory. This should help management in answering questions like (i) how often should the inventory system be monitored (ii) when should an order be placed and lastly (ii) what should be the order size.

1. Spare parts inventory may be held in several locations/companies such as the user of the equipment, the manufacturer of the equipment, regional warehouses, service parts center, or an organization whose business is to supply spare parts for repair and in some cases repair service. Therefore, stocking decisions at each location should be based on inventory investment, service levels, and technical capabilities. Generally speaking, slow moving and expensive items should be stocked at higher echelons provided the items can be shipped quickly. Conversely, high usage and inexpensive items should be stocked locally as close to the user as possible.
2. The ABC system for general inventory classification is also applicable to spare parts inventory management. It is one of the first steps in developing or reengineering an inventory management system. The premise of this analysis is that not all items are equally important, the greatest benefit can be achieved by focusing efforts on items of great importance (A items) and building appropriate slack in the system for categories B and C items to avoid problems.

   Different criteria can be used to classify the items in the three categories. The criterion most often used is the annual cost usage. However, in spare parts management, items can also be classified on the basis of criticality—items in category "A" being very critical while items in category "C" being less important. In spare parts management, criticality is closely related to the stock out costs (such as downtime cost of equipment when the part is not in stock). Factors other than annual cost usage and stock out costs can also be used to asses the criticality of the items. It is not unusual to consider factors like lead-time, obsolescence, engineering design volatility, and scarcity of product in the classification of items. Finally, one needs to establish different management controls (ordering policy and cycle count frequency) for each category commensurate with its importance.
3. Modeling the demand of slow moving items is an important first step for utilizing optimization models for inventory management. Parts fail for a variety of reasons such as normal wear, accidents, or misuse of equipment. Demand due to normal wear depends upon the age of

the products whereas in other cases it is a multitude of random factors. This makes determining the demand distribution a difficult task. Some of the distributions that can be used to model the slow moving items are Poisson, binomial, negative binomial, and beta. A distribution can also be fitted to a data set, and statistical tests can be performed to test for appropriate distribution. *Using the wrong distributions can lead to wrong policies such as incorrect determination of safety stocks and reorder points.*

4. Almost all analytical models use cost data to derive the inventory policies (such as order quantities and reorder points). It is therefore, important to estimate these costs reasonably well.
   - Holding costs or carrying costs is all the expenses incurred because of the amount of inventory carried. Usually, the following components are considered to be part of the holding cost:
     - Opportunity cost of capital (often turns out to be the most significant part of the holding cost).
     - Obsolescence, deterioration cost because inventory is no longer usable.
     - Cost of providing storage.
     - Taxes and insurance.
   - Shortage cost is the cost associated when the item is not in the stock. For spare parts, this is an important cost for it can be used to set the appropriate service levels. An increasing number of companies are implementing lean manufacturing techniques; manufacturing companies have thus become more vulnerable to outages. *Downtimes can be really expensive and in some cases life threatening.* In addition, companies are increasingly dependent on high-tech hardware and systems for business critical applications. For example, a simple stocking policy can be derived based on the expected marginal cost of carrying an extra unit in inventory versus the expected cost of stocking out.
   - Ordering cost is the cost of placing a purchase order with a supplier or the setup cost associated with ordering a manufactured lot from the plant.
5. Metrics are also important for service parts. Measures such as forecast accuracy, inventory turnover, obsolescence rate, fill rate, delay time, and downtime are some of the important measures that any maintenance department should keep track. Metrics can also be used to warn managers of mismanagement of inventory systems such as excessively high level of expensive items or excessive inventory levels for too many parts or too many stock outs.
6. A failure analysis is an important part of any forecasting system. It should include the failure rate for each part-machine combination,

the number of parts used in each machine, failure rates of each machine, and the number of installed machines. Thus, a good forecast should reflect the reliability of the parts and the installed base.

Appendix 2D Mobile Technology Case Studies*

Mobile solutions for *computerized maintenance management system* (CMMS) have been deployed within hundreds of companies worldwide, across industries. From heavily regulated industries such as pharmaceutical manufacturers and healthcare organizations to oil and gas or energy producers, to government agencies and corporate facilities, mobile computing has proven to be an effective means for improving productivity, lowering costs, and improving asset management programs. The following are just some real-life examples of mobile technology at work today.

Rush Medical Center, Chicago, Illinois

The largest medical facility in Chicago and one of the nation's largest, Rush Medical Center, has 85 technicians covering over 3-million ft^2 in a 28-building facility. The group averages 120,000 work orders per year, maintaining everything from plumbing and fixtures to critical high-tech medical equipment.

Rush Medical was challenged with a four-month work backlog, poor data collection, and a paper-based workflow that hindered productivity with lengthy end-of-shift paperwork. The resulting delays in information being inputted into the organization's CMMS kept the department in a constant reactive mode, trying to catch up while planned work further went unfinished.

Rush Medical implemented mobile solutions for work order management and maintenance inventory management. The organization deployed devices with docking cradles to preventive maintenance technicians and chose to have selected corrective maintenance technicians equipped with devices that could connect to the backend database wirelessly.

Within four months, Rush had already seen full project payback. They estimate an average savings of $1 million per year, and five years after deploying mobile, they are still generating benefits. The current 85 technicians are doing the workload of 108 by gaining 300 hs more of wrench time, which was previously spent on paperwork and excessive foot traffic. Planned maintenance compliance was up from 25 to 90 percent.

*These case studies are contributed by Erica LeBorgne, Syclo, Erica.LeBorgne@syclo.com.

The overall gains in efficiency helped reduce reliance on outside contractors to complete work requests, while patient satisfaction rose from 60 to 85 percent. Technician turnover disappeared, clerical workers were designated to other more important tasks, and inventory storage levels were significantly reduced, further helping to control costs. As a result of deploying mobile, the maintenance department reported a 28 percent increase in completed work orders.

Hewlett Packard, Corvallis, Oregon

Hewlett Packard's (HP) inkjet printer division in Corvallis, Oregon contains 11 buildings spanning 2.1-million ft^2. With more than 10,800 assets to maintain and a staff of 45 tradespeople, HP decided to cut operating costs and improve the life of assets by implementing a reliability centered maintenance strategy. The challenges were focused on maintaining asset uptime, because production line failure could cost HP up to $500,000 an hour. In order to better manage maintenance costs, the maintenance department had to increase productivity while maintaining the same staff.

HP deployed mobile devices that synched at the start of each shift and allowed workers to record progress while they worked. The result of mobile proved to be significant. HP estimates mobile drives $15M in potential quarterly savings from reduced downtime. Tradesmen have saved an average of 43 min per shift (equivalent to getting the work of five additional technicians per day without hiring any new people), which equated to a $4M annual savings for the company. The ratio of preventive maintenance to corrective work orders rose from 1:5 to an astounding 7:1. Maintenance costs were reduced overall by 25 percent, and HP was able to achieve project payback in just four months.

Columbia Gas Transmission (Eastern United States)

The operators of one of the largest storage, transmission, and distribution systems in the nation, Columbia Gas Transmission, supports several hundred gas end users in 11 states. With 13,000 mi of pipeline, over 100 compressor stations, 40 storage fields, and 3500 wells to maintain, the maintenance staff of 550 technicians and 30 supervisors needed the help of mobile technology to provide the visibility and information at the point of activity needed to maintain such a vast area of assets.

Columbia Gas faces unique challenges in providing timely reports for regulatory compliance, as well as audits from the various departments of transportation in their various states. The company also focused on efforts to eliminate technician travel time and data entry by removing paperwork from the equation, and to leverage the benefits of mobile to more quickly resolve emergencies. In addition, Columbia Gas was looking to mobile to complete more planned work and increase asset uptime.

Columbia Gas deployed a mobile work order management system to allow technicians to interact with their enterprise asset management (EAM) system from the road. They chose ruggedized devices from Itronix, the Husky, which were able to communicate via phone lines or local area networks deployed out in the field. Preparation time for audits was reduced from 19 to 28 weeks when using the paper-based workflow, to mere hours. The company gained in excess of 50 min per shift, per technician in productive time, equating to more than 100,000 hs per year. Field work orders increased by 30 percent, and the overall quality of life for technicians improved as paperwork and lengthy drives were eliminated.

Appendix 2E Mobile Technology Case Study*

Using Mobile Handheld Units in Conjunction with CMMS

TRW, a commercial steering gear manufacturer, has over 350 pieces of equipment that require frequent maintenance through the use of several types of lubricants, cutting oils, and other fluids. In order to track when and where the fluids are being used, the company implemented a *computerized maintenance management system* (CMMS) that resulted in over $15,000 in cost savings per year.

Prior to implementing maintenance software, TRW had no way of knowing when the fluids were used, how much quantity was used on each machine, and which equipment consumed more fluids than others. Employees completed a manual weekly inventory on paper to determine which items required reordering. This system provided a general understanding of the inventory shortage with a speculative cost estimate of total fluid use.

The company purchased new mobile units equipped with software that would capture transactions and input the information into CMMS to enable several types of analyses. The software shows different metrics such as fluid usage per item per month, usage per piece of equipment, and transactions per lube technician. The software also enables trend analysis, charting, and costing data from the collected information.

*This case study is contributed by Kathy Shoptaw, TRW Automotive, Kathy.shoptaw@trw.com.

The maintenance software eased the day-to-day operations and efficiencies for all the employees. Now, the lube technician scans the piece of equipment to which fluid is added, scans the part number of the fluid that is being added, scans their badge number, and types in the number of gallons of fluid added to the fluid reservoir. The entire process takes a few seconds, and at the end of each shift a synchronous coalescing is performed to capture all maintenance activities for that given shift.

The inventory table within the maintenance program enables management to access part numbers of the fluids that need to be reordered, analyze weekly costs, or other pertinent information. The data shows which machines use a particular fluid and how much is consumed, in addition, the maintenance software provides usage and costs per piece of equipment, per cell, per work center, and per department.

Further maintenance benefits include knowing which machines are leaking and which machines require repair. A proactive approach to maintenance helps reduce dollars spent on coolants and lubes.

TRW found significant cost savings and production benefits after implementing maintenance software. Using this automated process, the company reduced fluid usage approximately 10 to 12 percent in less than a year. This results in about $12,000 savings per year, and a total estimated cost reduction of approximately $15,000 per year. A computerized maintenance solution was absolutely successful and an integral part of TRW's overall fluid management program.

Chapter

3

How to Successfully Justify a Computerized Maintenance System

Introduction

Most managers find it increasingly difficult to control rising maintenance costs because of inadequate or outdated procedures. One tool that can help is a *computerized maintenance management system* (CMMS). The low cost of PCs and reasonable software cost put them within reach of many small maintenance shops. However, before considering the purchase of such a system, the company has to justify it, which basically means convincing people. How do you convince different levels of management?

First, maintenance managers must determine whether a CMMS is beneficial to their operations. They must ask questions such as:

- How long can the plant tolerate a production line breakdown due to part unavailability (if maintenance stores are not properly monitored, the company may face more costly breakdowns)?
- How much more are we spending on maintenance today than five years ago?
- Do we have the information we need to plan maintenance operations?
- Can we get this information when we need it?
- Is it in usable form?
- What are the company's plans for operations, for equipment?
- Will a computer really help?

The following questions also arise nearly every time maintenance is required on a piece of equipment:

- Where did we buy that last spare part?
- How much did we pay?
- Is this equipment under warranty?
- Who was the contact person we talked to?
- What was the phone number?
- Do we have a blanket purchase order with this vendor?
- How did we get the last part shipped?
- What was the delivery time?

Answers to these questions or inability to answer these will indicate a definite need for a CMMS. CMMS is useful even when the maintenance information system is basically sound but is not easily accessible or more information is needed.

Think about your maintenance system and how it affects other areas of operation such as production, accounting, payroll, and customer service. Could these operations be made more productive by improving the speed and efficiency of a maintenance information system? If the answer is yes, you could benefit from computerizing your maintenance system.

Another important issue to consider is the amount of information that can leave the company when a key maintenance employee leaves. Years of critical information can be lost the moment the employee walks out the door.

Roadblocks to CMMS Acquisition

Opposing the CMMS acquisition are the internal roadblocks that stand in the way of the system purchase. The following list will help you prepare for common roadblocks associated with acquiring a CMMS:

- *Budget not available now.* This is one of the most common excuses offered by management. It shows lack of acceptance and/or commitment by management.
- *Inadequate project payback or savings.* One must do a thorough job of determining benefits and savings. You will find CMMS very easy to justify in most cases.
- *Management information system (MIS) does not give high enough priority.* MIS does not give enough importance to a CMMS project, thus creating a roadblock. If MIS supports the project, its chances of success are increased greatly. CMMS is complicated to many decision

makers. Helping MIS understand the importance of CMMS should be a primary goal of every potential user. With MIS on your side, it is easier to convince others.

- *Failure to reach consensus.* All parties involved disagree on either the need for a CMMS or on the features required in a CMMS.
- *Company too small for a system.* This attitude suggests a basic lack of understanding of the true benefits and functions of a CMMS. It can pay for itself even for very small companies. There are many companies with just one maintenance technician successfully using a CMMS. A CMMS will help record and maintain the equipment histories that will be the basis for future repair/replacement decisions. An accurate and complete history can also describe how the job was done the last time, thus saving time associated with job redesign.
- *Prior attempt failed.* You should try again.
- *Do not have enough computer capability.* This is a common excuse. Computer hardware and software costs are all part of the justification process. Once the project is justified, purchase of computer hardware is not a problem.
- *Do not believe a CMMS will work in our situation.* As long as equipment requires maintenance, a CMMS will work.
- *Have never considered the benefits.* You should consider them now and go through the process of justification.

Step-by-Step Process

Following is an outline of procedures that can help you justify a computerized maintenance program for your application.

Form a team

The maintenance manager should assign one person (project leader) the responsibility of researching and justifying the CMMS.

Establish a team. The project leader should establish a team to assist him or her in the investigation. The team should consist of the plant engineer, maintenance manager, maintenance employees, and representatives from the data processing, purchasing, and accounting departments. Marketing, sales, and human resources should also be included. You should involve everyone who has any impact on this project. At a later date, the coordinator may also need advice from the company's legal department when agreements are needed between the software and hardware vendors.

Involving your employees in the automation process enables you to break down their resistance to computers and build enthusiasm for CMMS as a tool to facilitate their work.

Identify problems with present system

Problem analysis. First, determine exactly what problems pertain to the maintenance department. This is crucial to selecting and purchasing the proper CMMS package for your company's specific needs. Ultimately, we will determine the optimal system to solve (or minimize) these problems. For example, two very common problems that exist in companies today are

- Excessive downtime
- Lack of inventory control

Excessive downtime is a problem that occurs all too often for a variety of reasons. The causes range from lack of *preventive maintenance* (PM) to unavailability of parts when machines are down. The question that should be asked at this point is "can CMMS help with this?" The answer is "Yes." The properly selected CMMS package can help with PM scheduling and better inventory control.

Some problems simply may not be solved with a CMMS, such as employee theft.

Discuss and brainstorm with others in the maintenance department and other departments as well. Talk to anyone that you believe will benefit or be impacted from CMMS. Create a list of your own with as many descriptive problems as possible within and related to maintenance. Take time to consider the causes, not just the effects. Focus on all your maintenance problems.

Now compare your list with the following list and check to make sure relevant problems are covered. Feel free to add those to your list.

Reorganize your list until you have a final list by categories. Note: Some problems may belong to multiple categories.

List of problems (Example)

Labor productivity

1. Shortage of manpower
2. Shortage of crafts to finish job
3. Trouble justifying more manpower
4. Union problems
5. Employee—no means to monitor and control
6. Planning—material is not available when needed or too much is in stock
7. Planning—the proper tools are not available or accessible for the job

8. Planning—the necessary craftspeople are not available to do the job
9. Scheduling problems (manpower, material, equipment, and so forth)
10. Crisis management
11. Rescheduled job priorities
12. Job priorities—which to do first?
13. Utilization of manpower resources
14. Looking for supervisor to get the job assignment
15. Visiting job sites to determine what needs to be done
16. Rounding up materials and making multiple trips to the warehouse
17. Looking for tools
18. Waiting for other crafts to finish
19. Waiting for shutdown, clearance, and access to the job site
20. Time wasted due to lack of information or drawings
21. Time wasted due to canceled work orders
22. Waiting for special tools or engineering specifications
23. Paperwork—wasted time
24. Extended coffee or lunch breaks
25. Late start-up, early wash-up
26. Attitude and motivation
27. Lack of preventive maintenance
28. Paperwork—lack of storage space or simply unorganized

Equipment availability

29. Spare parts are out of stock
30. Justify upgrading existing equipment
31. Machine—no means to monitor and control/too much downtime
32. Planning—material is not available when needed to fix a machine problem
33. Crisis management
34. Lack of PM

Inventory control

35. Planning—material is not available when needed to fix a machine problem
36. The proper tools are not available or accessible for the job
37. Spare parts are out of stock
38. Too many parts
39. Too many obsolete parts
40. No information available for substitute parts

Product quality

41. Lack of history records
42. Lack of training
43. The necessary craftspeople are not available to do the job
44. The proper tools are not available or accessible for the job

Environment controls

45. Government/OSHA regulations
46. Safety procedures and standards

Management support

47. Management philosophy conflicts
48. Lack of communication with other departments/within the department
49. Lack of support from other departments
50. Lack of management support
51. Lack of long-term planning

Maintenance information

52. Assets—recordkeeping
53. Assets—identifying
54. Lack of history records
55. Cost control—budgeting/no control over budget

Determine objectives, features, and benefits

List short-range and long-range objectives. A typical short-range objective is to have inventory control computerized within 6 to 12 months. At that time you would have the facility for printing a current inventory status, reordering reports, and the like. Short-term objectives must complement the long-range objectives. Make a list of your long-range objectives over the next three to five years. Typical long-range objectives are to increase overall maintenance productivity and increase control through more timely information.

One cannot define good objectives without performing some preliminary work to determine these objectives. An excellent place to start is the list of your problems. For each problem definition there will be one or more objectives. For example:

Problem: Job planning time is high resulting in high job backlog; high emergency repairs.

Objective: Job planning time should decrease, backlog should decrease, and emergency repairs should decrease.

Put together a comprehensive list of your objectives. Some objectives (examples) are listed as follows:

Labor productivity

1. There should be sufficient manpower to carry out maintenance tasks.
2. There should be some way to measure and monitor employee productivity.

3. Material and tools should be available when jobs are scheduled.
4. Emergency repairs should be minimized.
5. Rescheduling of jobs should be minimized.
6. Work order instructions should be clear.
7. Equipment should be available when jobs are scheduled.

Equipment availability

8. Parts should be available when a job is scheduled for a piece of equipment.
9. Equipment replacement analysis should be performed to aid equipment replacement decisions.
10. Emergency repairs should be minimized.
11. Proper PM should be performed.

Inventory control

12. Parts should be available when needed and if appropriate.
13. Excess inventory should be minimized.
14. Obsolete parts should be eliminated.
15. Substitute parts information should be available.

Features. The next step is to compile a list of desired and needed features in your CMMS package. Based on your problem definition, you compiled a list of objectives. Now, based on each objective, prepare a list of features. For example:

Problem. Job planning time is high, resulting in high job backlog and high emergency repairs.

Objective. Job planning time should decrease, backlog and emergency repairs should decrease.

Features. Should have a work order scheduling module to help reduce job planning time and backlog. Should have a PM scheduling module to help reduce emergency repairs.

Now, put together a comprehensive list of features needed in your CMMS package.

The maintenance management system should be totally integrated and designed for use by all levels of personnel at the plant. The integrated system means information entered at one point will automatically and instantly become available at all other appropriate places throughout the system. The key sections of the program should include (examples):

- Equipment identification
- Referencing and cross-referencing
- Complete and comprehensive equipment specifications

- Spare parts lists for each piece of equipment
- Stockroom inventory levels and purchasing records
- PM worksheets and check lists
- PM schedules
- Work order preparation instructions
- Outstanding work order listings and priorities
- Maintenance histories and downtime records
- Manpower summaries
- Equipment listings
- Work order inquiry
- Equipment cost histories
- Major overhaul standards
- Uniform methods to initiate work
- Procedure to approve and authorize work
- Weekly review of work priorities
- Weekly work program based on craft net capacity to do work
- Uniform methods to measure and control work backlog
- Recording man hours, materials, and job costs
- Uniform procedure to report completed work

The system should provide maintenance personnel with a clear picture of what is expected of them and adequate feedback on their performance. The system should also interface with existing plant information system where possible and necessary.

Benefits. In deciding to obtain a CMMS, the motivation is usually a combination of improved production and a savings in maintenance time and costs. Savings and improvements take many forms. They may be large or small, but they will be realized if a CMMS is successful.

Your task is to put together a list of benefits that would be obtained for your organization if a CMMS were implemented. Start with your list of problems. For each problem, you defined one or more objectives. You documented features corresponding to those objectives. Now, prepare a list of benefits corresponding to the features. For example:

Problem. Job planning time is high, resulting in high job backlog and high emergency repairs.

Objective. Job planning time should decrease, backlog should decrease, and emergency repairs should decrease.

TABLE 3.1 Problems, Objectives, Features, and Benefits

Problems	Objective	Features	Benefit
Job planning time is high	Planning time per job should decrease	Proper WO scheduling module	Better scheduling: less time spent on planning
High backlog	Reduce backlog	Proper WO scheduling module	Reduce backlog: less time spent on backlogged projects.
Job scheduled - material not reserved, can't find parts, hence jobs are rescheduled	Material should be allocated when a job is scheduled	WO Module should allocate parts when WO is issued	Improvement due to better scheduling: less rescheduling of jobs
Parts not available, craft personnel either wait for new assignment, spend hours either locating parts or fabricating them	Parts should be available when needed	Inventory control module should have ROP feature	More productive time assured through parts availability
High emergency repairs	Minimized emergency repairs	PM module - PM available by calendar days & runtime	Reduce emergency repairs - increased machine output
Excess inventory	Reduce stock level	Inventory control module - ability to show low & high usage items	Reduction of stock levels - Reduction of carrying costs

Features. Should have a work order scheduling module to help reduce job planning time and backlog. Should have a PM scheduling module to help reduce emergency repairs.

Benefits. Increase in labor productivity (less money spent on overtime),

Better equipment availability resulting in higher capacity output.

This list of benefits is very important, as it will get translated into savings and help justify the system. At this point, you will end up with a spreadsheet like the one shown in Table 3.1.

Examples of benefits

Increased labor productivity. If the system provides the employees with a planned job, the procedures, needed parts and tools; the employees will be able to go directly to the job and do the needed work with no delays or interruptions. The employee will also work more safely, since job plans would include all safety procedures.

All this will increase labor productivity.

Efficient asset management and maintenance scheduling will contribute to reduced overtime. Improved parts tracking and availability will reduce the unproductive time of employees. Overall you will witness:

1. Reduction of overtime.
2. Reduction of outside contract work.

3. Reduced maintenance backlog.
4. Reduced cost per repair.
5. Improved morale of employees by diffusing employee frustration; a happy worker is a productive worker.
6. Better service to other departments.
7. A significant reduction in paperwork to make the most productive use of employee time.
8. Effective utilization of maintenance and supervisory personnel's time.
9. Reduced follow-up role required of the supervisor.

Increased equipment availability. Within a few months of implementation, it will become much easier to identify repetitive faults and trends. This information will assist in maximizing equipment uptime and reducing breakdowns. Your emphasis should shift from reactive to proactive maintenance, and your ratio of percent planned to unplanned jobs should increase. A corresponding reduction in downtime will follow.

Savings through reduced production loss results primarily from performing adequate *predictive maintenance* (PdM) and PM. PdM employs sensors or detectors to monitor equipment performance and condition. Detection of potential problems prompts the writing of the needed work order. For example, when an unacceptable reading of bearing temperature or vibration is sensed, a maintenance response can automatically initiate. The response may take the form of an automatic injection of lubricant or an initiation of a maintenance work order.

PdM helps keep the equipment in good condition. It leads to timely repairs rather than waiting for an actual failure to initiate corrective maintenance action.

PM is the regular scheduling of specific maintenance tasks to prevent possible anticipated failures. PM tasks include work such as a filter or bearing replacement, and calibration and condition checking.

An effective PM program will keep equipment in good condition because it forces periodic monitoring, and it serves as an early detection system for finding problems before they mature into full failures. The immediate result is that maintenance jobs are kept to minimum size. The long-term result is that equipment retains its effectiveness, value, and reliability in supporting production.

Longer useful life of equipment. You can prolong the effective lifetime of your assets and equipment through regular, adequate PM. The CMMS will support the processes involved in prolonging the life of your assets, which also improves resale value of the equipment.

Overall you will witness reduction of downtime.

Inventory control

Reduced Inventory Costs. Planning of jobs permits parts to be available when and where needed. Experience shows that a reduction of 10 to 15 percent in parts stocked and consumed is possible. Reductions also extend to inventory carrying charges, as well as to stockroom size, staff, and service requirements.

As work becomes more predictive so does your stock holding. Carrying out regular stock reviews allows you to minimize stock and to reduce expensive inventory. Spare parts can be linked to equipment, ensuring that obsolete parts are readily identified. Many users find that the greatest returns from a CMMS come through improvements in inventory control, with savings of 10 to 15 percent being typical.

Product quality. Improved product quality results primarily from performing adequate PdM and PM. PdM and PM details are described in the section "Equipment availability" earlier. Good equipment condition assures good product quality.

Environment control

Safety issues. Preventing accidents and injuries as a result of proper procedures documented by CMMS can save you a significant amount of money.

Compliance issues. Compliance with industry regulations. Some industries such as food processing, pharmaceuticals, and petrochemicals require that your asset management systems comply with the national or international standards that regulate their industry. If you require such compliance then you should select a CMMS that has a provision for the same. Meeting the regulatory requirements can save you money that you would otherwise pay in fines for not meeting the requirements.

Maintenance information. Access to maintenance information is dramatically expanded. Other benefits are also expected of a CMMS. Improved reporting and support for management control can contribute strongly to justifying a CMMS. Crucial production line decisions can be simplified by dependable and timely data on equipment condition, and expected lifetime. Such information can provide guidance in setting the size of production runs, deciding on equipment replacement, and pricing the product.

Better management of service contracts. Service contracts are arrangements with outside contractors for continuing services such as fork truck maintenance. Because these services are managed maintenance, it is convenient to use the CMMS to manage the functionality and accounting part of the services.

Blanket Purchase Orders. Blanket purchase orders with parts and materials suppliers may be managed effectively by a CMMS. This is an important area for tight control as blanket purchase orders are proven to be major leakages in the maintenance budget.

Increase in overall plant productivity. In the process of evaluating the benefits, you should remember not to look at CMMS as a tool to reduce employment. Instead, they should be viewed as a tool to reduce the rate of future hiring. For example, acquiring a CMMS may permit you to continue to operate with one maintenance clerk as your business expands, instead of hiring another one. The CMMS will, however, allow one maintenance clerk to get more work done and to do the job faster and more accurately.

Financial Analysis

Savings

The nature of savings through improved maintenance management can vary from one industry to another. Not all forms of savings and cost avoidance are available in every industry, nor are all savings sufficient to justify installation of a CMMS. Nevertheless, where justified, a system may pay for itself many times over through cost savings and avoidance.

So far you have determined the benefits derived from using a CMMS. Now, calculate savings offered in dollars by each benefit. You must do a detailed analysis of each benefit for your maintenance operation. For example:

Problem. Job planning time is high resulting in high job backlog and high emergency repairs.

Objective. Job planning time should decrease, backlog should decrease, and emergency repairs should decrease.

Features. Should have a work order scheduling module to help reduce job planning time and backlog. Should have a PM scheduling module to help reduce emergency repairs.

Benefits. Increase in labor productivity (less money spent on overtime), better equipment availability resulting in higher capacity output.

Labor productivity

First year savings = $45,000

Second year savings = $60,000

Third year savings = $60,000

Higher capacity output

First year savings = $120,000

Second year savings = $130,000

Third year savings = $130,000

TABLE 3.2: Savings From Using a CMMS (Blank Form)

Financial analysis: Savings			
Savings from using a CMMS			
Cost/year ($)			
Item	First	Second	Third
Totals =			

You can fill out Table 3.2 to document your savings. Put together as many benefit items as you have.

Table 3.3 shows typical savings generated in a maintenance department by use of a CMMS. If you do not have resources to compute detailed savings, you may use this data to determine savings for your organization.

Table 3.4 shows an example of savings as a result of using a CMMS.

A survey was published in the online magazine www.plantmaintenance.com, identifying areas of benefits from using a CMMS (see Table 3.5).

Cost estimates

Obtain rough cost estimates:

- Research websites, catalogs, publications, and technical literature
- Call on software suppliers

TABLE 3.3 Benefits/Savings Guidelines

	Range of savings, percent	Median, percent
Better scheduling Planning time per job decreases, resulting in more job plans produced, fewer backlogged jobs, and fewer breakdowns and emergency repairs. Job scheduling is more efficient through availability of reserve stores materials. Only jobs with materials available are scheduled, resulting in fewer schedule changes and reduced time spent by personnel waiting for materials. Job planning is improved by the systems support data, which include planner backlog and status report, quick recall of repetitious job plans, and computerized scheduling and historical job analysis.	5 to 12	8.5
Parts availability More productive time is assured through parts availability. When parts are not available, craft personnel not only idly await new assignments but also many times spend hours attempting to locate or even fabricate parts.	1 to 3	2
Machine availability Machine production time increases as the computer contributes to reduced emergency repair through the preventive maintenance program. The ability of the system to automatically schedule preventive maintenance puts this valuable program up front and removes it from neglect.	0.5 to 2	1.25
Stores inventory The ability of the system to maintain moment-by-moment inventory levels, automatic ordering, and parts cross referencing results in reduced inventory and fewer stockouts. The automatic cycle counts and updates inventory for efficient parts management.	10 to 20	15

TABLE 3.4 Example of Savings Using a CMMS

Savings from maintenance functions			
Labor productivity	Cost/year ($)		
• Improvement through better scheduling	Frist	Second	Third
• Improvements through parts availability	51,000	51,000	51,000
Stores inventory	12,000	12,000	12,000
• Reduction of stock levels			
• Reduction of carrying cost	67,500		
Machine availability	6,750	6,750	6,750
Increase in machine throughput			
	34,375	34,375	34,375
Totals:	171,625	104,125	104,125

TABLE 3.5 Survey Results Displaying Benefits From Using a CMMS

	Size of benefits obtained % of responses			
Area of benefit	Significant	Some	None	Don't know/not applicable
Reductions in labor costs	5.7%	32.4%	29.5%	32.4%
Reductions in materials costs	11.4%	32.4%	22.9%	33.3%
Reductions in other costs	8.6%	36.2%	23.8%	31.4%
Improved equipment availability	9.5%	37.1%	21.9%	31.4%
Improved equipment reliability	13.3%	41.0%	15.2%	30.5%
Improved cost control	35.2%	23.8%	16.2%	24.8%
Improved maintenance history	30.5%	37.1%	9.5%	22.9%
Improved maintenance planning	30.5%	36.2%	8.6%	24.3%
Improved maintenance scheduling	28.6%	39.0%	6.7%	25.7%
Improved maintenance schedules	29.5%	35.2%	9.5%	25.7%
Improved spare parts control	21.9%	35.2%	12.4%	30.5%

- Calculate total software costs:
  - System analysis and definition
  - Program installation and testing
  - Data entry (plant personnel or outside help)
  - Software acquisition cost
  - Installation cost
  - Cost of modification, if any
  - Training cost
  - Operating cost
  - Software maintenance cost
- Calculate total hardware costs:
  - Hardware
  - Maintenance
  - Supplies
- Calculate total cost

Include every possible cost factor.

Fill out Table 3.6. Add as many items as you can think of as part of CMMS implementation cost.

Determine net savings. Subtract total fixed cost from total savings to find net savings.

Compute return on investment (ROI). A ROI calculation for justifying a CMMS project results in a value that represents the savings (benefits) received from a CMMS against the total cost of implementing it.

TABLE 3.6 CMMS Implementation Cost Estimate (Blank Form)

Implementation cost estimate			
	Cost/year ($)		
Item	Frist ($)	Second ($)	Third ($)
Totals =			

The ROI can be calculated as follows:

$$\text{ROI\%} = \frac{\text{total savings} - \text{total costs}}{\text{total costs}} \times 100$$

Let us say, you are trying to compute the ROI for 2 years; then,

$$\text{Total savings} = \frac{\text{savings year 1} + \text{savings year 2}}{2}$$

$$\text{Total costs} = \frac{\text{costs year 1} + \text{costs year 2}}{2}$$

Example Table 3.7 shows an example of implementation cost estimate.

$$\text{Total savings} = 171{,}625 + 104{,}125 + 104{,}125 = 379{,}875$$

$$\text{Total cost} = 65{,}000 + 25{,}000 + 25{,}000 = 115{,}000$$

$$\text{ROI\%} = \left[\frac{(379{,}875 - 115{,}000)}{115{,}000}\right] \times 100 = 230\%$$

Set up Key Performance Indicators (KPIs)

Once the CMMS is implemented, you have to make sure the savings documented in justification are being realized. You also have to set KPIs

TABLE 3.7 Example of Implementation Cost Estimate

Implementation cost estimate			
	Cost/year ($)		
Item	First	Second	Third
Systems analysis and definition	10,000		
Software	5,000		
Software maintenance		500	500
Hardware	5,000		
Program installation and testing	10,000		
Data gathering and entry	15,000		
Operating cost	20,000	20,000	20,000
Totals:	75,000	20,500	20,500

in order to monitor and control the progress. You can define your own KPIs. For example, let us say you are operating at 40 percent labor efficiency and your goal is to increase that to 75 percent. Then your KPI is to obtain labor efficiency of 75 percent. You have to monitor progress and make sure your goal is achieved. You can revise your KPIs on an annual basis (or more frequently, if necessary).

Here are some examples of KPIs: (Goals indicated are typical industry examples. You can set your own goals.)

- *Maintenance cost per unit of production.* Obtained from completed work order CMMS reports and company financial reports. Benchmark actual costs against goals.
- *Percentage of planned work orders completed as scheduled.* To measure planning effectiveness. Obtained from CMMS for completed work orders, compare date required to date completed (goal: 95 percent).
- *Percentage of craft hours charged to work orders.* To monitor resource accountability: obtained from CMMS: report of actual craft hours charged to work orders (goal: 100 percent).
- *Percentage of planned versus emergency work orders.* To evaluate PM and planning effectiveness (goal: 90/10).
- *Percentage of craft utilization.* To monitor maintenance employees' productivity (goal: 85 percent).
- *Craft performance.* Actual versus estimated time. To monitor and maximize craft resources (goal: 95 percent).
- *Overall equipment effectiveness (OEE).* OEE is based on equipment availability, performance, and quality. This KPI is measurement of overall reliability improvement efforts (*goal: 85%*).
- *Percentage of equipment availability.* To monitor and minimize downtime (goal: 90%).

Conclusion

Follow the step-by-step process to justify a CMMS for your application. Identify your needs by reviewing your present practices. Determine objectives, features, and benefits of a CMMS. Compute total project cost, savings, and ROI. Once the system is justified, put together KPIs and monitor them on a regular basis.

Chapter

4

How to Specify, Evaluate, and Select a CMMS

Introduction

Once a *computerized maintenance management system* (CMMS) has been justified, the next step is to acquire a CMMS. Where do you start and how do you proceed? You can either develop the CMMS in-house or purchase ready-made software. Regardless of the option you choose, it is important to review a number of criteria discussed in this chapter to make the right selection for your application.

In-House Development

In-house development means software developed in-house by your own employees or people subcontracted to develop it for you under your direction per your specifications.

Ideally, an in-house system offers a great deal of flexibility to a company. It also provides the best link with existing plant information systems because it can be designed to accommodate the needs of other departmental systems. And, the in-house system can be designed to meet the highly specialized needs of a company, while readily adapting to its current maintenance operations.

Major advantages of in-house development:

- It provides the greatest amount of flexibility.
- It provides the best link with existing information systems.
- It can meet highly specialized needs.
- It has potential for optimum training development.
- It is readily adaptable to current maintenance operations.

Major disadvantages of in-house development:

- High cost
- Long development time
- Potential for failure to meet expectations: costs, target dates, scope, and system capabilities
- High potential for narrow focus in development rather than creative and innovative solutions
- Poorly documented in the rush to finish the job

Purchase Ready-Made Software

As indicated earlier, the in-house development alternative has a few serious problems. To quickly review, an in-house development could take a long time with a very high cost. With this in mind, it is time to look at purchasing a ready-made software.

Research has shown that purchasing software from a CMMS vendor can offer significant benefits to a company, especially in terms of saving time and money.

The reason for the significant savings in ready-made software is because you save many of the time consuming factors involved in in-house development. Problems like:

- Involving extensive programming efforts
- Extensive testing and redesign

Because the solutions are developed over a number of years, and designed to meet a variety of needs and specifications, they often prove to have more sophisticated, creative, and innovative approaches to maintenance solutions. These same approaches may never enter into the wildest dreams of an in-house programmer tasked with automating a specific system.

Advantages of ready-made software:

- Relatively low cost
- No development time
- Shorter implementation time
- Provides current state-of-the-art technology
- Provides additional features than your proposed system, thus further improving productivity
- Provides the required flexibility to meet your current and future needs
- User groups to help you get most out of your software

Disadvantages of ready-made software:

- It may not link with other existing information systems.
- It may not meet highly specialized needs.

This alternative brings up the questions: "How do I find and select the maintenance software that is best for my application?"

If a proper logical selection method is not followed, it may result in the acquisition of software that will not satisfy the company's needs, and involve delayed implementation, canceled projects, returned software, and frustration.

There are a number of ways to select software. You can rely solely on those programs that you see advertised in popular trade journals. You can call a competitor and ask him about the system he or she bought. You can hire a consultant to figure out what you would need, based on your requirements.

In an effort to achieve maximum results and return on your investment, you, the client, should manage the evaluation, acquisition and implementation process—from start to finish. This way, not only can you realize the potential benefit and drawbacks of a proposed system, but also you will be able to select the best package to suit your current and future needs.

Step-by-Step Process

No matter which way you choose, the following step-by-step process should be followed:

Form a team

The maintenance manager should assign one person (project leader) the responsibility of looking into evaluation and selection of CMMS suitable for your organization.

Establish a team

The project leader should establish a team to assist him in the investigation.

The team should be made up of the plant engineer, maintenance manager, maintenance employees, and representatives from the management information systems (MIS), purchasing, and accounting departments. Marketing, sales, and human resources should also be included. You should get every one involved that has any impact on or of this project. At a later date, the coordinator may also need advice from the company's legal department when agreements must be made and CMMS vendor may be an excellent add-on to the team.

By getting your employees involved in the automation process, you break down their resistance to computerization.

Determine the objectives

The first step is to develop a complete outline of the scope, goals, and objectives of the proposed system. Determining your objectives was discussed in Chap. 3, justification; let us expand on that here.

The following steps will assure that adequate attention is given to the first and most important phase of the project:

1. Identify the major problems to be solved by the implementation of CMMS. As a double check on the real priority of these items, roughly quantify the benefits these improvements will yield. How much money can be saved by achieving the selected objectives? Defining current system problems will provide management with evidence of the need to resolve them. It is important that problems be well documented in terms of lost revenue, inefficiency of operation, and inability to meet current and future business goals.

   The major benefits will vary from plant to plant, industry to industry, and organization to organization. In one situation downtime on production equipment may be the primary concern, whereas in another it may be the productivity of the maintenance personnel. The system should be specified to achieve the maximum economic advantages in an individual situation.

   Once your problems have been identified and defined, objectives for the proposed system can be set. As an example, let us say one of the problems is excessive inventory. The objective simply would be to reduce inventory.

   Once the objectives are set, you can determine features required in your CMMS based on the objectives. In the example above, the feature would be:

   The CMMS package should have an inventory control module with ROP capability and usage history monitoring and reporting capabilities.

2. In the development of the objectives for the new system, input should be obtained from every organization and level that will interact with it. All levels of maintenance management should have an opportunity to make their needs known and make their contributions.

   Unless the system being scoped is strictly internal to a department, there should be input from other departments. Production personnel should be consulted about proposed CMMS since they are almost always the primary customers. Other groups that may have an interest and be able to make contributions to the planning phase include purchasing, receiving, cost accounting, and payroll.

3. It is important that the system being considered is not limited to automated version of existing procedures that might prevent exploring additional features, which might further improve maintenance productivity.

The following steps are recommended to ensure that new ideas are being input into the planning process:

Review trade journals. Thoroughly review current trade journals for new ideas that are working in other similar plants and organizations. Some of the trade journals you may wish to look at include Maintenance Technology, Plant Engineering, and Plant Services.

Review available product literature. Gather software product information to review the features offered by available systems. By exploring a wide range of systems, you open doors to broaden your range of possible solutions. The product literature should provide you with enough information to help you decide whether or not the package merits your further attention. Do not hesitate to review the features offered by the largest and most sophisticated CMMS package for good ideas.

The question is where do you obtain the list of CMMS vendors? A good place to start is related trade journals as mentioned. Internet (web) search is another place to obtain a lot of information regarding CMMS vendors. Some other avenues are outlined as follows.

Attend seminars. Attend seminars covering the latest development in CMMS. Look for practical, hands-on workshops and seminars. List of seminars can be found at several places. Professional organizations and Internet are good places to start.

Contact a consulting firm. Discuss your situation with reputable consultants and other experts in the field of modern CMMS.

Be sure the consultant has no financial interest in your selection. Remember, you want clean and clear information to be the determining factor in any decision you make. Be certain the consultant you choose has an in-depth understanding of your maintenance situation, as well as a thorough understanding of existing software and new developments.

Contact other companies. Contacting noncompeting firms in other industries that employ systems in similar applications to your own can prove quite fruitful.

Identify the hardware alternatives

The ideal situation is to match the hardware to the most suitable CMMS package available. Normally, this degree of freedom is not available to the maintenance department. Company policy may dictate use of certain hardware.

It is strongly recommended that you work as closely as possible with the MIS group. Listen carefully to their recommendations, especially with regard to technical matters. MIS, like maintenance, is a service organization trying very hard to keep their customers happy in a rapidly changing environment.

Develop the system specifications

With objectives and a general outline of the system agreed upon, the next step is to define, in detail, the specific features desired of the new system.

1. Identify the major aspects of the system.
2. Clearly define all significant features.
3. Group the features within each subdivision into three categories of importance:

   Mandatory. Features of highest importance. Absence of one or more of these would make the system unacceptable.

   Needed. Features of great value. The omission of a few scattered items does not indicate outright rejection, but reduces the value of the system.

   Desired. Features of marginal value. These are features you would like to have, but can do without.

When you begin to categorize and prioritize, make sure you consider your long-range plans. What may appear as desired or needed features now, may prove to be mandatory as your company grows.

Values then can be assigned to each feature on a scale of 1 to 10. Mandatory features would automatically be assigned a value of 10. It is suggested needed features be ranked 4 to 9, and desired features from 1 to 3.

At this stage, you should have a document outlining all the features required in your CMMS. This document called "system specifications" will be submitted to software vendors later to obtain pricing, and so on.

Appendices 4A and 4B illustrate examples of system specifications.

In App. 4A, the company has put together a very detailed set of specifications. This takes a fair amount of effort and resources; however, it eliminates many undesirable responses. Appendix 4B shows a brief, but to the point specifications.

Search for CMMS vendors. By now, you probably have a good list of vendors already to work with. If you wish to do further research and add more vendors to your list, now is a good time. How many vendors do you start with depends on availability of time resource. It is recommended you start with no more than 10 to12 vendors.

Conduct preliminary screening

When the initial system's literature has been gathered, a quick preliminary review should be made as follows:

1. List the system name, vendor, and compatible hardware.
2. Quickly review the details to determine how many of the mandatory features appear to be provided by each of the candidate systems.
3. Eliminate systems that do not meet the majority of the mandatory features.

Further evaluation

Review the top three or four candidates with the following criteria in mind:

System features

Flexibility. The software should be flexible in terms of allowing you to enter information pertaining to your organization. It should accommodate your current as well as future needs.

The CMMS package should allow you to start small and then add additional areas when needs develop. For example, you might start with a work order system and add a purchase order module at a later date.

Limitations. Check the limitations of the system. For example, if number of records in the database increase significantly in future, the system should not slow down search and reporting capabilities. You do not want to invest a great deal of time and effort in a software program that will not work after a short period of time. (If the software presents limitations, ask the vendor if it could be modified to overcome them.)

Interfacing capabilities. The system should be capable of interfacing with other information systems. For example, you might want to interface the CMMS with your existing purchasing program. The CMMS should provide data export and import capability.

Self-sufficiency. Programs should be capable of direct, full use without or with minimal need to consult a manual or other outside sources. Instructions on screen should explain what the program will do and how to use it.

Operating system. Software should support the most current operating systems. For example, MS Windows XP.

Programming language. Normally, you will not have a choice of programming languages. However, if there is a difference in programming language between competing products, the programming language used may be a factor of your choice. A program written in one language may perform better than another. This can be a significant factor if you are dealing with thousands of records.

System security. Security is a very important function of a CMMS package. A multilevel password protected security system can define exactly what each user is permitted to access, edit, or run. A good security system allows the following:

- Make different screens for different users (e.g., a person just making a work request would view a very simple work order screen compared to the maintenance supervisor who will view the whole work order screen).
- Make field visible or hidden (you do not want everyone to view employees' pay rates).
- Make fields editable or noneditable (this provides database security).
- Make fields mandatory.
- Validate data entry against predefined criteria.

Data security. The program should provide some backup facility to protect against data lost through accident or otherwise.

User customizable reports. This option allows you to customize reports without any programming knowledge. You are not dependent on the vendor for report modifications that saves you time and money. With this option, you can:

- Modify reports and forms
- Create new reports just the way you want them
- Interface with other programs such as accounting, purchasing, even if they are running on a different operating system
- Create ASCII file formats for universal data import/export

User customizable screens. This option allows you to customize screens without any programming knowledge. You are not dependent on the vendor for screen modifications that saves you time and money. With this option, you can:

- Modify a screen
- Change name (legend) of a field

- Change the size of a field
- Change the position of a field
- Add or delete fields
- Modify context sensitive help (this is very important since you are allowed to modify fields. If you modify a field and cannot change corresponding help, it could create major problems.)

Modifications. Each company operates a little bit differently. A very sophisticated feature in the software for one company may be a limitation for another. The vendor should be able to make the custom modifications to suit your application. If you are going to invest the money, it might as well be for something that you really want and is going to work for you. (It becomes more significant as the majority of the vendors do not supply the source code.) In many instances, using the software for three to six months provides you with the opportunity to go into fine details of the program. You can identify possible enhancements in the program that will further improve your productivity. At this point, you may be approaching the vendor to explore the feasibility of making those modifications. That is why it is important for you to make sure the modification service is available.

From those vendors who do not provide a copy of source code, a written contract should be obtained stating that a copy of the source code will be kept with an attorney or some other escrow agent, and the copy of the source code will be made available to the user in the event something happens to the vendor (going out of business, filing bankruptcy, and the like).

User ease

Easy to learn

Training. Software suppliers provide training aids for their software in a variety of ways. Training aids include workbooks, on-line tutorials, videos, CDs, DVDs, and web-based training. Make sure training aids are easy to understand and well illustrated.

Documentation. The software should be accompanied by an instruction manual. As a matter of fact, it is highly recommended to review the instruction manual before buying the software. Although a well-designed, easy-to-use software eliminates the need for a manual, it is important to have one in case you get stuck somewhere.

Easy to use. While the system must be easy to learn, it must also be easy to use. Once you have evaluated how easy the proposed system is to learn, it is necessary to review it in terms of day-to-day operation.

Menu driven. The software has to be very easy to use without the aid of manuals, and so on, as much as possible. A good package is icon and menu driven. You just keep clicking on icons or menu items for navigation and to perform desired CMMS functions.

Input screens. Input screens should be simple and uncrowded. Information entered in the system should be placed on the same spot on the screen, as it will appear on the paper forms you use to collect the data. If there are default values, they should automatically appear.

Error handling. Check if there is good documentation of error codes? Does the program "trap" user errors well, that is, does it catch your mistakes and tell you how to correct them?

Context-sensitive help. The CMMS should have context-sensitive help. Wherever you have a question in the application, if you seek help, it will guide you through that part.

Vendor profile

Qualifications. Check the qualifications of the vendor staff. What kind of experience and background do they have pertaining to maintenance management and CMMS? Remember, you want to deal with people who are knowledgeable in the field, who can offer you the help and assistance you need to improve your maintenance operation. There are probably close to 300 CMMS vendors. Every year certain percentage of companies go out of business and new ones are formed. Many of these are headed by people with programming or sales background with no one on board who has maintenance and CMMS background.

Financial strength. Check vendor's financial strength. How stable and established is the company? The CMMS project is a major investment in terms of time, resources, and money. Your relationship with your vendor could extend for quite some time.

References. Telephone interviews, followed by email exchange if appropriate, should be conducted with a number of users of each vendor's system to determine their experience and satisfaction with the package.

1. Prepare an outline of the information to be gathered and the specific questions to be asked for each user.
2. Obtain a comprehensive list of users from each vendor; preferably a list of all users, not a select list of a few of the vendor's favorites.
3. In making your selection of users to contact, follow the vendor's suggestions for two or three, and then pick several more on your own. Perhaps these would be from your industry or companies of your own size. Visit some of the users if possible.
4. Try to talk to at least two different people at each location to minimize individual biases. It is desirable to talk with both a system user, and a person in MIS who supports the system.

Following are some of the suggested questions:

- How long have you had the system?
- How long did it take to implement?
- How effective was the training?
- How has vendor support been?
- Would you recommend this software, and the vendor?
- Why was this package chosen?
- How can it be improved?
- What benefits have been realized since the system was implemented?

Delivery. Is the package available now? Do not depend on a package that is supposed to be available on a certain date. Computer program expected completion dates have been known to be very inaccurate, even by giant software companies.

Terms and conditions

Payment. The payment terms may weigh heavily on the final decision as it affects your cash flow. Typically, most vendors ask for a certain amount as down payment, a certain amount upon delivery, and balance upon acceptance within a certain time period. For example:

10 percent with purchase order

80 percent upon delivery

10 percent upon acceptance (within 60 days)

Source code. Usually, the software vendors do not give a copy of the source code to clients. An acceptable method is placing a copy of the source code with an independent third-party escrow agent. This is necessary to protect the vendor's proprietary rights and the client's interest.

As a system user, failing to have the source code, or at least access to it, could pose some serious risks, should something happen to the vendor:

- The vendor may become financially unsound; as a result, the source code could be tied up in a bankruptcy proceeding, or confiscated by creditors.
- The vendor may become the victim of a fire, or some disaster, which destroys the source code.
- The vendor may simply fail to perform. Perhaps they changed the focus of their business and no longer provide support.

Warranties. Software vendors generally try to overprotect themselves by limiting liabilities. Aside from the guarantees cited, warranties are largely symbolic as far as a software buyer is concerned. Given the complexity of many programs and wide variety of uses by thousands of users, it is difficult to see how they could be otherwise.

The software should be guaranteed to perform per published specifications. The vendor should guarantee to fix errors found at no charge within reasonable time frame.

Vendor support

Demonstration. Check to see if the vendor has a provision for demonstrating the software. In most cases, you should be able to download the software demo from their Web site. Some of them might mail you a CD. Make sure it is a fully functional demo. You can arrange personal meeting with the vendor after narrowing the list down to the finalists.

The idea of trying the software before making the final purchase is to make sure it is going to work for you.

Training. Check if the vendor has a provision for training, either at their facility or on site. It is always a good idea to go through a training session provided by the vendor. In the long run, the small investment will save you a great deal of money and frustration.

Investigate the following

- Training cost
- Number of training days recommended
- Training location

System support. Vendor support for your system is very important. Check to make sure they have the necessary technical ability to answer your questions. You may need extensive customer support during early stages of implementation. Support should include:

- A telephone response line
- E-mail set up for quick response to your questions
- A newsletter that provides advice and tips on how to get more out of the system. It should contain information on latest upgrades to the software.

Upgrade policy. A progressive software company periodically upgrades their products based on feedback received from users and potential users. Check to see if the vendor has an upgrading policy? If so, what kind? Typically, annual maintenance covers all future upgrades.

System cost. This is a major criterion. You should investigate all possible cost factors associated with the software, such as:

1. Software acquisition cost
2. Up-front costs (sales tax, delivery, and so forth)
3. Documentation cost
4. Implementation cost (installation, and the like)
5. Training cost
6. Travel and other expenses
7. Maintenance cost (ongoing) (*Typically 10 to 20 percent of the software cost, per year*)
8. Customization cost

Make sure there are no hidden costs involved. Try to obtain a price guarantee clause stating that the vendor must guarantee its price for 120 days or so from the date the proposal is submitted.

Compile, Compare, and Select

With all the steps of this evaluation procedure completed, you are ready to compile the results, make comparison, and the final selection. See Fig. 4.1.

Follow the model illustrated in Fig. 4.1

Select the vendor that provides the best combination of characteristics for your particular situation. If two or more systems are approximately equal, so much the better. Now enter into serious discussions with the vendors and attempt to eliminate the weakest characteristics of each of these systems through negotiation. It may be good to ask for assistance from your purchasing department and MIS. Often they can play the "tough guy" role by demanding price or technical concessions while you, as the user, maintain a smooth relationship with the potential vendor.

Appendix 4C shows a comparison of some CMMS vendors.* You can make a matrix of your own and use these features to compare vendors.

Conclusion

A properly selected CMMS will grow with your business as data accumulates in the database. It will maintain its value for a long time, resulting in a reduced total cost of ownership. It will also enhance the productivity of your users and add profits to the bottom line of your organization.

*Source: Plant Services, www.plantservices.com/cmms review.

Features	Rating	Vendor A		Vendor B		Vendor C	
	Rating	Rating	Rating	Rating	Rating	Rating	Rating
	I	A	I × A	B	I × B	C	I × C
General							
1 Easy to use (menu driven)	9	8	72	7	63	7	63
2 Interface capability	7	6	42	0	0	9	63
3 Multi-level password system	9	7	63	5	45	9	81
4 Back-up utility	7	2	14	9	63	6	42
5 Archive capability	8	7	56	8	64	8	64
6 Graphics capability	5	9	45	2	10	7	35
Inventory control							
1 Automatic reorder	9	8	72	8	72	8	72
2 Cycle count supported	5	0	0	6	30	7	35
3 ABC Analysis	5	0	0	0	0	5	25
4 Barcode support	8	0	0	8	64	9	72
5 Imaging support	9	8	72	8	72	8	72
6 Multiple vendors/part nos.	9	7	63	9	81	8	72
7 Multiple locations	10	7	70	8	80	9	90
Workorders							
1 Adhoc search	9	5	45	7	63	8	72
2 Work history	9	8	72	8	72	8	72
3 Parts cost tracking	10	7	70	9	90	9	90
4 Provision for outside costs	8	5	40	7	56	9	72
5 Failure codes	8	2	16	5	40	8	64
6 Parts allocation	9	7	63	9	81	9	81
7 Overtime tracking	7	5	35	9	63	7	49
Total =			910		1109		1286

Figure 4.1 Final software selection.

Appendix 4A System Specifications Example 1*

Introduction and instructions to suppliers

Introduction. We invite suppliers to submit proposals for the *computerized maintenance management software* (CMMS) project in accordance with the requirements, terms, and conditions of this *request for proposal* (RFP).

This RFP sets forth the requirements for all products and services and solicits a detailed response from suppliers to include pricing and service descriptions in the specified format.

Main objectives include, but are not limited to, the following.

*This appendix is contributed by Jerry Eaton, Mercury Marine, Jerry_Eaton@mercmarine.com.

Communication process. We have instituted a cross-functional team to develop an overall plan that enables "company" to improve efficiency and achieve better tracking of our overall maintenance operations with respect to *preventive maintenance* (PM) and *work orders* (WO), on a corporate-wide basis. The team has prepared the RFP, and will evaluate the supplier proposals. As a part of our process, we require that all communication regarding this RFP, and the associated effort, must only be directed XXX and will be coordinated internally among our team, as required. We will consider any supplier attempts to circumvent our process a breach of that process, which may result in termination of that supplier from further consideration for this business expenditure.

General conditions. This RFP is not an offer to contract. Acceptance of a proposal neither commits our company to award a contract to any supplier, even if all requirements stated in this RFP are met, nor limits our right to negotiate in our best interest. We reserve the right to contract with a supplier for reasons other than lowest price.

Failure to answer any question in this RFP may subject the proposal to disqualification. Failure to meet a qualification or requirement will not necessarily subject a proposal to disqualification.

We may choose to include any portion of your response in an agreement that may be a result of this RFP.

Valid period of offer. The pricing, terms, and conditions stated in your response must remain valid for 180 days from the completion of the proof of concept phase.

Confidentiality/nondisclosure. The information contained in this RFP (or accumulated through other written or verbal communication) is confidential. The information is to be used by each supplier only for the purpose of preparing a response to this RFP. The information may not be used or shared with other parties for any purpose without written permission from Mercury Marine.

Right of rejection. We reserve the right to accept or reject any or all responses to this RFP and to enter into discussions and/or negotiations with one or more qualified suppliers at the same time, if such action is in the best interest of our company.

Cost of proposals. Any expenses incurred in the preparation of proposals in response to this RFP are the supplier's sole responsibility.

Proposal delivery. Deliver one electronic mail copy of your proposal to the following address, not later than 5 pm CST on January 23, 2003.

Company information

RFP questions. Questions regarding this RFP are encouraged and should be submitted in writing via electronic mail to e-mail address.

Questions shall be received no later than January 21, 2003. All questions will be answered in writing within two working days. Suppliers are required to supply in their request, current physical and e-mail address in order to ensure prompt delivery. Answers to questions from any supplier will be provided for review to all suppliers.

Schedule of events. See Table 4A.1.

Customer profile

Company information

Executive summary. Identify the key products and services you are proposing in response to our RFP requirements. Provide a high-level discussion of the overall costs as well as benefits we will experience from each product/service that your company will provide. Provide a general overview of any services or functions your company agrees to include at no charge in order to enhance the overall value of your proposed product/service package. This discussion should be geared toward our senior executives, and focus on the quantifiable benefits of your proposal.

Corporate profile

Company background. Please provide a brief overview and history of your company. Include any information on proposed or planned mergers or partnerships your company is involved in. Describe the organization of your company and include an organizational chart.

Financial information. Provide audited financial information on your company (e.g., annual report, 10-K).

Reference accounts. List three accounts similar in size and/or complexity to our organization that we may contact. Please include the following information:

TABLE 4A.1 Schedule of Events

Dates	Activity
31 Jan. 2003	RFP distribution
7 Feb. 2003	Questions deadline
14 Feb. 2003	Proposals due before 5:00 pm CST

- Company names and address
- Contact name and phone number
- Products or services provided to these accounts

Please provide contact information for two customers that were not completely satisfied with your software and/or support.

General questions

"Company" requests that each supplier provides a complete proposal based on the information contained in this RFP. Provide full explanations when requested, or as necessary to further describe your product/service. If you do not comply completely with any statement as written, please indicate and then comment how you deviate. Also provide the date that any product/service will be available, as appropriate. You may add additional separate explanations as long as the order of your response is in continuity with this RFP. Any response left blank will be considered as a negative response from the supplier. If a question does not apply then put "does not apply" and explain why. Please respond to each of the following questions:

1. Where are your corporate offices? Where are your branch office locations (city, country, and so forth)? Describe the functions of your branch offices.
2. State your current number of customers and where "company" would rank in your customer list (i.e., number 25 out of 200)? Who are your three largest customers? Include the number of concurrent users for each of these customers.
3. What industry trends are you particularly sensitive to? How are you positioning your organization to respond to these trends?
4. Who are your top competitors? What do you consider to be their biggest weakness? What are their strengths? What distinguishes you from your competition?
5. Support team structure. As our business has continued to grow and change, "company" has come to rely heavily on support teams. With the increasing workload placed on our in-house staff, it becomes vital that your organization possess the skills and desires necessary to bring value to the arrangement. These teams must display efficiency and assertiveness in supporting our in-house staff. The team members must also show ownership of their area of support. Please provide us with a list of staffing and resources that will be assigned to manage and support our relationship. Please include any executive support and escalation procedures.

6. Research and development—please explain how much money has been spent on R&D in the last three years; forecast for the next three years.
7. How does your company stay in touch with the needs of the market place?
8. Future capabilities—what are your plans for future capabilities? Discuss the next three revisions/upgrades to your software and what are the planned release dates.
9. Describe your options and recommendations for training of our organization.
10. Service and support—what levels of technical support can you provide and what are their costs? What is available through the web or telephone? What are your typical response times to problems? What constitutes a service call directly to software engineers (second level support) versus going through your help desk (first level support)? How do you measure service levels and response times met by your company?
11. How do you schedule releases of upgrades and patches? How do you deploy your upgrade and patches? Are the costs of upgrades and patches included in your standard maintenance contract?
12. Interfaces—what will be provided to interface your software with our current legacy system and Oracle Suite? How do you determine which and how many consultants will be assigned to work with Mercury through the pilot, point of purchase, and implementation phases? Will the same person/people be available through the entire project?
13. Based on company's requirements, propose a detailed implementation plan including all requirements for: (a) a pilot system and (b) a production system. These plans should include software and services required for these two phases. Please include a *work breakdown structure* (WBS) in MS project.
14. Are there any modules or additions to your software that your company offers that is not included in this proposal?
15. How is the concurrent user volumes determined for optimal usage?
16. Suggest particular features of your package that should be proven out in a pilot? Base this on your past experience of other implementations.
17. What CAD viewers have been tested with your software? Can current "company" CAD viewers (Voloview and Product View) be proven in the pilot?
18. What are company's responsibilities in performing a pilot and implementing your software?

19. Describe how your software is licensed—concurrent users or registered user? Include in your discussion what constitutes a concurrent user and/or registered user.
20. Describe other pricing models available by which your software is licensed—enterprise, server-based, and so on.
21. Does your software use an internal or external report writer for generating reports, explain. Can other report writers such as discover and business objects be used?

Costing and pricing

Discounts and prices

Discounts. Provide pricing for products, licensed software, and services. Company pays on a Net Prox 25th basis from date of invoice receipt.

Prices. Please read and familiarize yourselves with the requirements. Please itemize all prices, discounts, and charges. Any taxes or duties are to be shown separately. Cost-saving alternative proposals are welcomed.

Include a complete description and pricing for any services/options your company offers (training, technical support, maintenance fees, third party software, and the like). Please itemize all prices, discounts, and charges. Any taxes or duties are to be shown separately.

Cost structure. The software provider will install and configure the software on company's servers. The software will be tested as outlined later to ensure all requirements are verified (proof of concept pilot) at software provider's risk and at no cost or obligation to company. The proof of concept will take a minimum of 90 days. No cost will be incurred by company during this phase. After completing the proof of concept (pilot) phase, the company will determine if the software should be purchased for full implementation. Company will be under no obligation to purchase the software regardless of the success of the proof of concept. See Table 4A.2.

System functional specification

Table user rollout. Table 4A.3 assumes that the proof of concept (pilot) phase was successful to warrant rolling the solution out to all "company" plants in North America.

Please review our requirements listed later and state if your software performs these functions line item by line item (yes or no). An explanation for each line may be added if it is needed for clarification.

TABLE 4A.2 Cost Structure

Implementation			Line cost
	Core software		
	Service request module (unlimited)		
	Database conversion		
	Interfaces: (Accounts/payable, human resources, fixed assets, general ledger, tru-trak purchasing requisitions, supplier master, pims, constraint planning mfg. software)		
	Training		
	Total users (1,000)		
		Total	
Support			
	Yearly support	Total	

Requirements

A. Software requirements

a. Contain at a minimum the following "modules"

1. Request for service
2. Equipment/assets
3. Work orders
4. Requisitions/*purchase orders*(PO) (Purchasing)
5. Project management
6. Preventive maintenance
7. Labor
8. Crib inventory
9. Bar coding
10. Scheduling/planning
11. Vendors
12. Interface
13. Report generator

b. Additional requirements

1. Divisions/business units/locations/plants/departments (hierarchy of at least five levels)
2. Capable of managing multiple databases world wide from one server

TABLE 4A.3 User Rollout

Phase	Estimated users	
	(New)	(Total)
Pilot	25	25
2003	375	400
2004	600	1,000

3. Capable of multiple factory/facility sites
4. Extensive security protection
5. With 2000 users no degradation in speed of the software

B. IT requirements
   a. Three-tier Internet-based architecture that includes (1) client, (2) application server, and (3) database server.
      1. Client
         a. Capable of running on a Microsoft Internet Explorer Version 5.5 or higher
         b. Capable of running on Windows 2000 workstation or higher
         c. Ability to run on low-end workstation of 128 MB memory, 233 MHz, and 4GB hard-drive
         d. Runs on standard workstation of 256 MB memory, 1.8GHz, and 20 GB hard-drive
      2. Application server
         a. The application server should be capable of running on Red Hat Linux operating system with Tomcat or Apache Web server.
         b. Capable of running on Oracle 9IAS application server is a plus.
         c. Ability to interface with LDAP compliant security directory.
         d. The application should be written to be compliant with Java J2EE standards.
         e. Access to the database should be made through Oracle SQLNET or ODBC.
      3. Database server
         a. Database server should be Oracle 8.1.7.4 or higher.
         b. Electronic messaging and interfaces should be made through FTP, SQL calls, or send mail to Lotus notes.
         c. Should run on TCP/IP network.

C. Company's minimum requirements
   a. Request for service module (hotline)
      1. Populates field automatically based on a few entries: employee's name/telephone number/location
      2. E-mail compatible
      3. 100 percent web-based
   b. Equipment/asset module
      1. Interface with Oracle suites-asset database
      2. Transfer PIMS asset database to Oracle database
      3. The ability to assign internal asset numbers
      4. The ability to assign special priority codes for extremely vital equipment
   c. Work order (WO) module
      1. Track machinery asset uptime/downtime.

2. View comprehensive and detailed planning information: schedules, costs, labor, materials, equipment, failure analysis, and so forth.
3. Ability to punch on and off WO.
4. Ability to incorporate bar code and PDA technology.
5. Ability to generate WO numbers by plant and year.
6. Assign cost centers.
7. Ability to manage multiple personnel punched on to one WO.
8. Track machinery uptime/downtime.
9. Ability to create duplicate WO from a master WO.
10. Ability to prioritize WO.
11. Ability to pull parts, reserve parts.
12. Track cost and taxes.
13. Enter work request on the spot without equipment details or background data from any Internet capable workstation.
14. Schedule WOs based on real-time critically and logistics.
15. Compare real-time budgets or estimates against actual and historical data.
16. Track monthly budget estimates against actual maintenance cost.
17. Track monthly budgets and actual for each cost center.
18. Notification of WO assignment to workers/supervisors/managers.
19. Task library of common maintenance work instructions.
20. The ability to set up multiple shifts including fifth and sixth shifts.
21. Uneditable history files.
22. Maintain and view a complete history of revisions.
23. Sign off/approval:
    a. Set a time period when a sign off should be made, and if that time has exceeded, the system would automatically escalate it to someone else.
    b. Configure sign off workflow to bypass or reassign a sign off due to vacation or create backup personnel.
    c. Ability to have sign off of up to three signatures.
    d. View the status of a revision sign off.
    e. Create and edit a sign off work flow.
24. Ability to prioritize WO.
25. Interface with accounting systems—allocate cost to cost centers.
26. Interface with Oracle Suite-GL/assets.
27. Pull in information from PIMS regarding old material history.
28. The ability to attach photo/drawings/text/and the like to WO.
29. Full-query function including query of description field.
30. Pull in PIMS asset database.

31. Maintenance/plant engineering personnel must be able to write WO.
32. Ability to generate both canned and customizable reports.
33. Ability to prioritize asset.

d. Requisitions/POs module
   1. Manage requisitions for both stock and nonstock items
   2. Include all costing, vendor, and quantity information when requisitioning
   3. Generate automatic replenishment on the basis or reorder parameters
   4. Interface with current requisition and purchasing systems

e. Project management module
   1. Track capital and expense data
   2. Integrate project work with the WO module
   3. Ability to assign project categories (safety, energy, HVAC, and the like)
   4. Ability to import vendor information
   5. Project planning
   6. Time lines (milestone, meeting minutes)
   7. Track RFPs and quotes.
   8. Track actual, budgeted, and committed cost
   9. Import and maintain CAD files, drawings, photos, and the like
   10. Inventory tracking
   11. Track and sign off tracking of all facets (HVAC, electrical, plumbing, fire, CAD, and so on) of the project
   12. If under project, ensure all requisition are tied to project number
   13. The ability to customize project numbers (such as year/project number—2003–0001)
   14. Maintain a full audit trail of all approval activity
   15. Ability to sort by plant/category/year/project manager/cost
   16. Ability to track project status—project cost, open/close date, milestones

f. PM module
   1. Access flexible online scheduling options for planners, schedulers, supervisors, and technicians.
   2. Import/Export PM/WO to production scheduling (Mercury's I-2 system).
   3. Define planning periods.
   4. Facilitate maintenance scheduling by viewing PMs scheduled days, weeks, or months in advance.
   5. Autoschedule (days, cycles, holidays, shutdowns).
   6. Load balance manpower.

7. Build crib kit and place in reserve (track under crib inventory).
8. Multiple PM task trigger points (time, meter, variable, and the like).
9. Routing to include up to three signatures.
10. PM either fixed interval or float based upon last completion date.
11. Forecast labor, parts, and tool requirements.
12. Suspend schedules when equipment is moved, is placed out of service or retired.
13. User-defined fields.

g. Labor module
   1. Create employee to include basic employee data, multiple labor rates and crafts, and training.
   2. Maintain personnel files for each employee's attendance, vacation, sick, and nonproductive work time; track overtime history, and individual pay rate(s).
   3. Establish regular and over time pay rates by crafts (up to five per individual).
   4. Organize labor by craft to streamline work assignment process.
   5. Track address, phone, fax, pager, e-mail, Web address.
   6. Track supervisor, department, division.
   7. Set nonworking days per employee
   8. Ensure that an individual cannot punch on to more than one WO at a time

h. Crib (nonmanufacturing) inventory module
   1. Ability to handle multiple cribs—definable by user.
   2. Manage crib request including vendor performance.
   3. Track warranty.
   4. Track stocked/nonstocked, taxable/nontaxable, rebuilt, substitute, and superceded.
   5. Track part specs (size, shelf life).
   6. Status reports and notification of WO/PM awaiting parts.
   7. Track quantity on hand.
   8. Multiple vendors per part, multiple order options per vendor.
   9. Multiple manufactures per part.
   10. Reconciliation with requisitions, POs, receipts, invoices, and so forth.
   11. Track quantity allocated to active WOs.
   12. Track quantity reserved for PM based on schedule.
   13. Track maximum/minimum ordering.
   14. Part stock and shortage tracking.
   15. Automatic requisition generation.
   16. Inventory adjustment tracking including tracking reason for adjustment.

17. Check in/check out.
18. History log.
19. Ability to handle bar code enabled counting.
20. Issuing and handling.
21. Electronic notification of arrival.
22. Backorder and partial shipment processing.
23. Repairable spare tracking.
24. Crib transfers.
25. Return to supplier accountability.
26. Identification of disposal of obsolete or inactive materials.
27. Return of used/unused WO/PM materials.
28. Stock forecasting.
29. Inventory valuation.
30. Ability to view crib inventory history by part number.
31. Allow multiple cribs within a plant.
32. Allow the same item within multiple cribs.
33. Allow the same item to have different attributes (i.e., reorder parts) within multiple cribs.
34. The ability to generate a receipt for incoming items.
35. Track withdrawals, order receipts, and cycle times.
36. Identify hazardous materials.
37. The ability to track items by serial number.
38. Track items with prints and revisions of prints. Signal changes.
39. Track items that can be "reused" or reconditioned.
40. Flag system—example the withdrawal/usage of this part is higher than normal.
41. The ability to assign a serial number to a part for tracking its movement such as an individual circuit board.
42. The ability to track warranty work by bar code.
43. Track changes in prices.
44. Track types of items in crib (tooling, maintenance, gauging, and so forth).
45. Allow new and used prices and quantity tracking.
46. Track last withdrawal date and department.
47. Ability to cycle count—automatic generation of daily cycle counts based on ABC class and user-defined time periods.
48. Allow purchasing and issuing in different units of measure.
49. Automatic suggestions of reorder points.
50. Allow reordering based on reorder points and lead times.
51. Allow ABC classification.
52. Allow tracking of tax.
53. Allow multiple locations within a crib.
54. Allow items to be purged while maintaining audit trail.
55. Archive data schemes.

56. Allow visibility to items on order and in transit.
57. Allow issuing to an account number (interface GL) ordering from asset numbers.
58. Allow interfacing to purchasing system (allow for blanket orders and spot buys).
59. Allow for interfacing to vendor master.
60. Allow for partial receipts.
61. Allow by over/under receiving.
62. Allow automatic suggestions for reorder points based on usage.
63. Allow receiving by part or by requisition.
64. Allow receiving without an order number.
65. Track individual(s) who withdrew items by item number and individual.
66. Receive and withdraw by both bar code and manually.
67. Ability to print a receiving slip from the receiving screen.
68. Crib requirement must be sent to PIMS/CPMS to manage budgets as well as to generate orders from the purchasing system.
69. Need security on update screens.
70. The ability to archive crib part numbers that have not seen activity over a period of time and still have access to the data if needed.
71. Online help screens.
72. The ability for crib managers to add text to requisitions.
73. Notification to "save" or "not complete" if you try to move to another screen without finishing a task.
74. Reporting:
    1. Top 50 items of on-hand dollar value by crib and by item type
    2. Complete inventory list by vendor
    3. Withdrawal aging—Example: Generate a report for all items not withdrawn for one year
    4. Past due report (items with past due orders)
    5. Query for items that are already held in inventory (by description or maker)
    6. "Mega data screen" all status and inventory history by part number
    7. Need to view data on screen to add/remove columns, or change format prior to launching the report
    8. Ability to print only the first, last, or any screen within the report
    9. Ability to export data from reports to other users or software packages such as Excel
    10. The ability to print screen for hard copy

i. Bar coding/PDA module
   1. The software must have the ability to use bar coding and PDA.

j. Scheduling/planning module
   1. Access flexible online scheduling options for planners, schedulers, supervisors, and technicians
   2. Import/export PM/WO to production scheduling (Mercury's i2 system)
   3. Define planning periods
   4. Automatically or manually enter schedules for variable period lengths
   5. Easily assign and plan shutdown periods and insert into regular planning periods using user-defined parameters
   6. Determine availability by the number of craft hours for a craft and plan.
   7. Utilize backlog management for resource leveling
   8. Automatically match task to the talent or skill of a worker to certain job
   9. Use various forecasting reports to assist in planning and analyzing availability for PM work, special project to be completed
   10. Automatically or manually enter schedules for variable period

k. Vendors module
   1. Track address, vendor number, active/inactive, phone number.
   2. This module will interface with our current vendor database.

l. Interface module
   1. Ability to interface with Oracle GL, Oracle HR, Tru-Trak, Oracle fixed assets, accounts payable, purchasing requests, vendors, production constraint management software (i2) and PIMS.

m. Report generator module
   1. Standard reports library
   2. Unlimited user-defined selection filters
   3. Customizable reports

D. Pilot requirements—"Company" will develop test scripts based on proposed business scenarios to validate the following functions:
   1. WO/PM: Tracking, approval process, material/repair history tracking, more than one punched-on, machine uptime/down time, scheduling, multiple labor rates, allocation to cost center, interface with crib withdrawal system, prioritization, and so forth.
   2. Report generator including ad hoc reports.
   3. Describe your internal security structure for the application (i.e., security for restrictions by plant, and user terminal).
   4. User friendly—drag and drop and pull down menus.
   5. The ability to make changes to the core software.
   6. Create a sign off workflow to include more than five signatures.

7. Monitor network response times.
8. Track material history of assets.
9. Execute the WO/PM that has multiple approval workflow routings (up to three signatures, flexibility to customize, bypass or reassign the sign off due to vacation or absence, escalation as needed, and so on).
10. Process the sign off workflow and view the status through the process.
11. Verify final flowchart to ensure routing of information is correct.
12. Forward a message from the application into Lotus notes and e-mail it.
13. Call software provider's help and tech support desk. Monitor response time.
14. Search software provider's support Web page.

Appendix 4B System Specifications Example 2

CMMS Specifications/Demonstrations/Review

Module 1: functional location or hierarchy of equipment

The module demonstrates how equipment locations are developed and how equipment specifications within those locations are documented. In addition, various links should be discussed.

Demonstrate the following

- Build a minimum of three equipment locations, each containing at least several specific pieces of equipment within those locations. Example: Number1 Paper Machine Silo Fan Pump is an equipment location. In this location are a motor, pump, coupling, electrical breaker, and electrical controls.
- The assignment of accounting designations (e.g., cost centers and department numbers).
- If a facility has an equipment location hierarchy (numbering system for equipment locations) already established, demonstrate how this can be linked or reestablished with the new software. Example: 611-01-01 is an established service location number for number1 Paper Machine Silo Fan Pump.
- Once an abundance of equipment locations are built, demonstrate the number of ways one can search and find equipment and specific information about that piece of equipment.

- Links attached to equipment locations that might include Safety/ZES text, drawing information, ISA Loop information, and engineering design specifications.

Module 2: equipment specifications and parts lists

The module demonstrates how equipment and associated parts lists are identified within locations.

Demonstrate the following

- How specific information about a piece of equipment is documented. Build equipment specifications for a motor, pump, and vessel.
- How specific pieces of equipment are identified with equipment numbers.
- The ability to search for equipment based on MFR name, location, specifications, and the like.
- Once a wide range of equipment is established, demonstrate abilities to analyze equipment inventory.
- Ability to print various types of data based on equipment specifications.
- How pieces of equipment are moved around within locations (equipment change-out).
- Build spare parts lists for equipment.
- Linking drawings, procedures to location.
- Ability to build a "common" *bill of material* (BOM) for equipment. Example: A number of Goulds Pumps could all have similar parts like bearings and seals. Demonstrate how one BOM can be made that links to all these pumps where one up-date to the common BOM updates all BOMs.
- Ability to build a "specific" BOM for equipment. Example: The same pump may have a special impellor. In addition to the common BOM, demonstrate the ability to create a specific BOM linked to equipment.
- Ability to search for parts linked to BOMs.

Module 3: work requests, work orders, and work order planning

The module demonstrates how work orders are written and progress from a notification stage through work order completion. Also includes a discussion of work order planning.

Demonstrate the following

- Ability to search and find a variety of equipment to which a work request can be written. (e.g., write a W/R to fix a bearing in a pump, patch a hole in a floor, and troubleshoot a system malfunction)
- Ability to code the work request with a number of "flags" (e.g., safety, regulatory, priority, identifiers, and outage related)
- Ability to write work requests to multiple crafts and how multiple craft requests are managed.
- Ability to have an "approve" function so work requests become "approved" work orders and how that functionality is managed. Demonstrate an example.
- Ability to "plan" a work order and access a "preplanned" work order. Demonstrate an actual planning example of work that requires a work plan, materials to be ordered, material available in stores, coding work order for scheduling, man-hour estimation, cost estimates, special equipment needs, safety text, priority setting, and other documentation that can be attached to plan.
- Ability to recognize "status" changes in work order progression (needs approval, approved but not planned, in planning, on schedule, and so forth).
- Ability to interact with existing purchasing system to provide BOM links to stores inventory, online requisitions, online catalogs.
- Ability to search for work orders within a backlog by multiple codes or descriptions within the work request/work order.
- Ability to print hard copies of work orders and other various forms of attached documentation.
- Ability to create daily and weekly work order schedules for single or multiple crafts that may include job descriptions, hours, manpower, "pop-up" preventative routes that are due, and critical path considerations.
- Schedule effectiveness or reporting ability to optimize scheduling function.
- Ability to export data to MS project.
- Ability to print "hardcopies" of work orders demonstrating the available information provided, attachments, and special instructions.
- Ability to "close-out" a work order demonstrating the history recording functionality, how history codes are used, how additional documentation can be stored and the overall general process of managing work order backlogs for history retention. Demonstrate closing multiple work orders.

- Access to a comprehensive work order file maintenance function allowing details within the work order to be changed and modified easily.
- Ability to access "real-time" cost data regarding materials, purchase orders, regular and overtime labor charges and other associated costs. Discuss facility requirements to accomplish a "real-time" system. Demonstrate tools and reports to manage costs with a focus on managing labor costs.

Module 4: preventive maintenance (PM) routes

The module demonstrates how PM Routes are created and managed.

Demonstrate the following

- Create several PM routes for various applications. One might include a motor lubrication route for 10 motors in multiple locations performed every six weeks. Another might be a more specific inspection of a gearbox performed annually. Describe "copy" or "autobuild" functionality.
- Options for creating permanent route numbers or explain route-numbering functionality.
- Ability to create routes with specific instructions, customizable text, data-links, graphics, trendable data, and the like.
- Route frequency coding, "grace period" functions, and completion/calendar-based logic that programs a route to come due on a desired frequency.
- How individuals would manage their PM backlog by knowing what is due, when the PM can be done, how the routes may populate into the scheduling function, and so on.
- Ability to close a PM route with history, enter PM route comments, write work orders from PM close-out, and create comments on future PM routes from past PM routes.
- Ability to view or print PM history comments, PM activity, due versus completion reporting, and other various reporting functions populated from the PM module.
- Record lubrication requirements (type, frequency, and quantity).

Module 5: history

The module demonstrates how equipment history is created, managed, and utilized.

Demonstrate the following

- Ability to create history via work orders, PM routes, stand-alone history comments, and other forms of history. Include how these comments may be linked to closed work orders, PM Routes, and cost data.
- Ability to view and print history data for specific equipment or systems, or by other varieties of codes (like dates, ID numbers, departments, crews, and W/O identifiers), which would include history comments, cost data, linked documents, searches resulting from history codes, and other forms of useful data.
- Ability to assist RCM efforts by providing information on equipment with excessive/repetitive failures and equipment failure reports.
- Reporting abilities resulting from historical data, which may include daily, weekly, or monthly (or other chosen frequencies) reports of work complete, failure analysis, cost data, or other useful information.

Module 6: central functions

The module demonstrates how the software can link data from multiple facilities.

Demonstrate the following

- Ability to view and copy data from other facilities like equipment specification, text, planned work, and equipment spare parts.

Module 7: overall system functionality

The module demonstrates some overall qualities of the software.

Demonstrate the following

- Overall user friendliness of software, discuss transaction speed, ability to easily move from procedure to procedure, use of multiple windows, error message quality, printing capabilities, use of communication tools like email, and so forth.
- Use of current technology, which may include access to vendor databases, graphics and graphing, use of PDAs or other devices, voice entry of data, connection to Internet links, and the like.
- Ability to provide "online" and "customizable" help instructions and examples.
- Ability to support root cause failure analysis, enhance basic care and involvement of operators with their equipment rounds, mean time

between failures (MTBF), troubleshooting database, and other support functions that increase equipment reliability.

- Ability to transfer data from existing software to new software. Data may include equipment specifications, equipment spare parts lists, equipment history, PM routes, functional location codes (service locations), and other forms of data.
- Ability to provide training to various levels of software users.
- Ability to provide training on best practices that enhance equipment reliability (like planning, RCFA, and basic care).

Appendix 4C CMMS Vendor Comparison*

Question Keywords			
Company	Ashcom	Champs	Datastream Systems
URL	www.ashcomtech.com	www.champsinc.com	www.datastream.net
Package Name/ Version	MaintiMizer 3.5 Build 103	Champs CMMS/EAM 8.3	Datastream 7i 7.6
Review Date	2/28/2005	2/17/2005	2/18/2005
Number of Customers	3000	76	7345
Pricing	$1900 for single PC standalone system, $1400 per concurrent user for five or more	$2700 per concurrent user, $50,000 minimum for software/services	$7999 per concurrent user to $2599 per user
Hosting Available?			$38 per month per user
Rental Available?			$170 per month per user
Hosted Database	N/A	N/A	Single database for all sites within a given company
Single/Mix	N/A	N/A	Single
Maintenance fee	20%	17%	18–20%
E-procurement	No	No	Yes
E-marketplace	No	No	Yes
Data Exchange	No	No	Yes
Newsgroup	Yes	Yes	Yes
Language	Visual C++	PowerBuilder	Oracle Forms, PLSQL, Java

*Plant services, www.plantservices.com/cmms review.

Company—package-price

Question Keywords			
Company	IFS North America	Indus International	Ivara
URL	www.ifsna.com	www.indus.com	www.ivara.com
Package Name/ Version	IFS applications 2003	Indus asset suite 5.0	Ivara EAM 4.0
Review Date	2/18/2004	2/23/2005	1/28/2004
Number of Customers	2800	>350	>50
Pricing	$30,000 base plus $500–$2000 per user	$125,000 plus $1500–3000 per user	$5000 per concurrent user
Hosting Available?		Yes	
Rental Available?			Yes
Hosted Database	N/A	Single database for all sites within a given company	N/A
Single/Mix	N/A	Single	N/A
Maintenance Fee	18%	18–24%	18%
E-procurement	No	No	No
E-marketplace	No	No	No
Data Exchange	No	Yes	No
Newsgroup	No	Yes	No
Language	Java with PLSQL and C++	Java	C++

Company—package-price

Question Keywords			
Company	MicroWest Software	MRO Software	SAP
URL	www.microwest software.com	www.mro.com	www.sap.com
Package Name/ Version	AMMS 10	Maximo enterprise suite 6.0	mySAP business suite and R/3 4.7
Review Date	12/192003	2/25/2005	2/18/2004
Number of Customers	>1800	10000	>2500
Pricing	$995 per concurrent user	Small/medium business: $15,000 per group of 5 users Enterprise: $175,000 for 50 users, $3550 per each additional	$1000–$3500 per user
Hosting Available?		$200+ per month per user	Yes
Rental Available?	No		
Hosted Database	N/A	Single database for all sites within a given company	All hosted sites and companies are on a single database with appropriate security between companies
Single/Mix	N/A	Single	Mix
Maintenance fee	12%	15–20%	17%
E-procurement	No	Yes	Yes
E-marketplace	No	No	No
Data Exchange	No	Yes	Yes
Newsgroup	No	Yes	No
Language	C#, MS visual studio.NET	JSP, Java (J2EE-compliant)	ABAP, C++, Java

Integration

Question Keywords			
Company	Ashcom	Champs	Datastream Systems
Cost % Hardware	2		
Cost % Software	35	30	50
Cost % Customization	10	10	10
Cost % Business Redesign	3	8	
Cost % Implementation	50	52	40
% Implementations Targeting KPIs	15		95
KPI No.1	Total cost of ownership		Inventory
KPI No.2	Inventory level		Unplanned downtime
KPI No.3	Parts availability		Warranty claims
KPI No.4	Labor performance to standard		Completed PMs
KPI No.5	Regulatory compliance		Production uptime
Cost Savings (%)			Not tracked
Revenue Increase (%)			Not tracked
Shared Fee (%)			
Shared Rewards (%)			
No Incentive (%)	100	100	95
Tracking	No formal survey	Yes, but results are not published	Yes, results are available upon request
Third-Party ERP	Yes	Yes	Yes

Question Keywords			
Company	Ashcom	Champs	Datastream Systems
Third-Party CAD	Yes	No	Yes
Third-Party PdM	No	No	Yes
Third-Party Data	No	Yes	Yes
Third-Party Handheld	Yes	No	Yes
Other Integration		No	
Phase-In	Yes, very graduated steps from simple to comprehensive system, priced incrementally	Yes, very graduated steps from simple to comprehensive system, priced incrementally	Yes, very graduated steps from simple to comprehensive system, priced incrementally
Remote db Synch	No	No	Yes
Browser Look and Feel	Yes, but browser version is limited in terms of functionality and/or look and feel	Yes, but browser version is limited in terms of functionality and/or look and feel	No client/server version available

Integration

Question Keywords			
Company	IFS North America	Indus International	Ivara
Cost % Hardware			0–5
Cost % Software	60	40	40–50
Cost % Customization			0–10
Cost % Business Redesign		20	10–20
Cost % Implementation	40	40	25–40
% Implementations Targeting KPIs	10	80	100
KPI No.1	Uptime		MTBF
KPI No.2	Inventory turns		Mtce contributed unavailability
KPI No.3	OEE		Mtce cost
KPI No.4	Life cycle cost		Percentage proactive mtce
KPI No.5	Mtce per unit production		Percentage planned mtce
Cost Savings (%)	Not tracked	20	5–25
Revenue Increase (%)	Not tracked	30	5–10
Shared Fee (%)			0–5
Shared Rewards			
No incentive (%)	100	100	95–100
Tracking	Yes, results are available upon request	Yes, but results are not published	Yes, but results are not published
Third-Party ERP	Yes	Yes	Yes

Question Keywords			
Company	IFS North America	Indus International	Ivara
Third-Party CAD	Yes	Yes	Yes
Third-Party PdM	Yes	Yes	Yes
Third-Party Data	Yes	Yes	Yes
Third-Party Handheld	Yes	Yes	Yes
Other Integration		GIS/mapping and E-procurement	
Phase-In	Yes, very graduated steps from simple to comprehensive system, priced incrementally	Yes, as above but no incremental pricing	Yes, as above but no incremental pricing
Remote db Synch	Yes	Yes	No
Browser Look and Feel	Yes, same functionality and same look and feel	No client/server version available	Yes, but browser version is limited in terms of functionality and/or look and feel

Integration

	Question Keywords		
Company	MicroWest Software	MRO Software	SAP
Cost % Hardware	5		
Cost % Software	70	60	40–60
Cost % Customization	5	10	0–10
Cost % Business Redesign	5		0–10
Cost % Implementation	15	30	20–30
% Implementations Targeting KPIs	10	50	75
KPI No.1	Downtime	Equipment availability	Planned/ unplanned ratio
KPI No.2	Planned/unplanned ratio	Labor productivity	Equipment uptime
KPI No.3	Cost per PO	Inventory cost	Preventive/ breakdown/ planned labor
KPI No.4	Stock outages	Unplanned downtime	Maintenance cost/unit production
KPI No.5	Inventory costs	OEE	Inventory level
Cost Savings (%)	20–30	20–30	
Revenue Increase (%)		10–20	
Shared Fee (%)	2		
Shared Rewards (%)	2		
No Incentive (%)	98	100	100
Tracking	Yes, but results are not published	Yes, and results are made public	Yes, but results are not published
Third-Party ERP	Yes	Yes	Yes

Question Keywords			
Company	MicroWest Software	MRO Software	SAP
Third-Party CAD	Yes	Yes	Yes
Third-Party PdM	Yes	Yes	Yes
Third-Party Data	Yes	Yes	Yes
Third-Party Handheld	Yes	Yes	Yes
Other Integration		Any system	RFID, GIS, route planning
Phase-In	Yes, very graduated steps from simple to comprehensive system, priced incrementally	Yes, very graduated steps from simple to comprehensive system, priced incrementally	Some flexibility
Remote db Synch	No	Yes	Yes
Browser Look and Feel	Yes, but browser version is limited in terms of functionality and/or look and feel	No client/server version available	Yes, same functionality and same look and feel

Basic capabilities

Question Keywords			
Company	Ashcom	Champs	Datastream Systems
Voice	No IVR capability	No IVR capability	Some
Format Error-Check	Some	Comprehensive	Some
Range Error-Check	Limited	Comprehensive	Some
Logic Error-Check	Limited	Limited	Comprehensive
Error Message Quality	Somewhat detailed	Somewhat detailed	Somewhat detailed
Custom Error Messages	No	Some	Some
User-Defined Defaults	Some	Some	Some
Multilanguage Log-On	No	No	Yes
MultiLanguage Toggle	No	No	Yes
Quick Data Entry	Some	Limited	Comprehensive
Online Help	Limited	Some	Some
Help Examples	A few	Some	A few
Help Tailoring	Same help for all users	Same help for all users	No
Training Days, Supervisor	2		2
Training Days, Planner	3		4
Training Days, Tradesperson	2		1
Training Days, Buyer	1		2
Custom Business Rules	By any user	Power user only	Power user only
Workflow Integration	Some	Some	Comprehensive
Drill-Down	Some	Some	Some
Editing	View only	Some	Comprehensive

Basic capabilities

Question Keywords			
Company	IFS North America	Indus International	Ivara
Voice	Comprehensive	Some	Some
Format Error-Check	Some	Some	Some
Range Error-Check	Some	Some	Some
Logic Error-Check	Some	Some	Limited
Error Message Quality	Somewhat detailed	Somewhat detailed	Somewhat detailed
Custom Error Messages	Yes (all)	Yes (all)	Some
User-Defined Defaults	Some	Some	Some
Multilanguage Log-On	Yes	No	Yes
Multilanguage Toggle	Yes	No	No
Quick Data Entry	Comprehensive	Comprehensive	Comprehensive
Online Help	Comprehensive	Comprehensive	Some
Help Examples	Some	Some	Some
Help Tailoring	Somewhat	Very much	Somewhat
Training Days, Supervisor	3	3	$^1/_2$–2
Training Days, Planner	5	6	1–3
Training Days, Tradesperson	$^1/_2$–1	2	$^1/_2$–1
Training Days, Buyer	5	5	1
Custom Business Rules	By any user	Power user only	By any user
Workflow Integration	Some	Some	Some
Drill-Down	Some	Comprehensive	Comprehensive
Editing	Comprehensive	View only	Comprehensive

Basic capabilities

Question Keywords			
Company	MicroWest Software	MRO Software	SAP
Voice	No IVR capability	No IVR capability	No IVR capability
Format Error-Check	Some	Some	Comprehensive
Range Error-Check	Some	Comprehensive	Comprehensive
Logic Error-Check	Limited	Limited	Comprehensive
Error Message Quality	Somewhat detailed	Somewhat detailed	Comprehensive
Custom Error Messages	Yes (all)	Yes (all)	Some
User-Defined Defaults	Some	Some	Some
Multilanguage Log-On	No	Yes	Yes
Multilanguage Toggle	No	No	No
Quick Data Entry	Comprehensive	Some	Some
Online Help	Some	Some	Some
Help Examples	Extensively	Some	Extensively
Help Tailoring	Somewhat	No	Somewhat
Training Days, Supervisor	2	1–3	5
Training Days, Planner	3	1–3	5
Training Days, Tradesperson	1	1–3	2
Training Days, Buyer	2	1	2
Custom Business Rules	Vendor only	By any user	Power user only
Workflow Integration	Some	Comprehensive	Comprehensive
Drill-Down	Some	Some	Some
Editing	Comprehensive	Some	Comprehensive

Work order scheduling

Question Keywords			
Company	Ashcom	Champs	Datastream Systems
Work Order Hierarchy	Yes (1–2 levels)	Yes (5 or more)	Yes (5 or more)
Blanket Work Orders	No	No	No
Time-Based Priorities	1	1	1
Criticality Priorities		1	1
Health and Safety Priorities		1	1
Other Priority	Equipment risk and condition		
User-Defined Priorities		Unlimited	Unlimited
Multiple Priorities	No	Yes	No
Schedule Display	No	No	Yes
Graphical Scheduling	No	Some (third-party S/W)	Comprehensive
Schedule Drill-Down	No, not from within scheduling	Yes, including work order edit capability while scheduling	Yes, including work order capability while scheduling
What-If Scheduling, Priority-Based	No	No	Yes (third-party S/W)
What-If Scheduling, Duration-Based	No	Yes (within CMMS)	Yes (third-party S/W)
What-If Scheduling, Labor-Based	No	Yes (within CMMS)	Yes (third-party S/W)

SOURCE: Plant Services, www.plantservices.com/cmms_review

Work order scheduling

Question Keywords			
Company	IFS North America	Indus International	Ivara
Work Order Hierarchy	Yes (5 or more)	Yes (5 or more)	Yes (5 or more)
Blanket Work Orders	No	Yes	No
Time-Based Priorities	1	2	1
Criticality Priorities	1	1	1
Health and Safety Priorities			
Other Priority	Critical %		Scheduling
User-Defined Priorities	Unlimited	Unlimited	Unlimited
Multiple Priorities	No	No	Yes
Schedule Display	Yes	No	Yes
Graphical Scheduling	Some (within CMMS)	Some (third-party S/W)	Comprehensive
Schedule Drill-Down	Yes, but work order information is display only	Yes, including work order edit capability while scheduling	Yes, including work order edit capability while scheduling
What-If Scheduling, Priority-Based	No	Yes (third-party S/W)	Yes (third-party S/W)
What-If Scheduling, Duration-Based	Yes (within CMMS)	Yes (third-party S/W)	Yes (third-party S/W)
What-If Scheduling, Labor-Based	No	Yes (third-party S/W)	Yes (third-party S/W)

Work order scheduling

Question Keywords			
Company	MicroWest Software	MRO Software	SAP
Work Order Hierarchy	Yes (1–2 levels)	Yes (5 or more)	Yes (5 or more)
Blanket Work Orders	No	Yes	Yes
Time-Based Priorities	1	1	1
Criticality Priorities	1	1	1
Health and Safety Priorities			
Other Priority		Location	
User-Defined Priorities		Unlimited	>10
Multiple Priorities	Yes	Yes	No
Schedule Display	Yes	Yes	Yes
Graphical Scheduling	Some (within CMMS)	Some (within CMMS)	Comprehensive
Schedule Drill-Down	Yes, but work order information is display only	Yes, including work order edit capability while scheduling	Yes, including work order edit capability while scheduling
What-If Scheduling, Priority-Based	Yes (third-party S/W)	Yes (third-party S/W)	No
What-If Scheduling, Duration-Based	Yes (third-party S/W)	Yes (third-party S/W)	Yes (within CMMS)
What-If Scheduling, Labor-Based	Yes (third-party S/W)	Yes (third-party S/W)	Yes (within CMMS)

Work order control

Question Keywords			
Company	Ashcom	Champs	Datastream Systems
Resource Scheduling	Comprehensive	Some	Comprehensive
Resource Analysis	Some	Some	Comprehensive
Resource Costing	No	Some	Comprehensive
Part Source Identification	Limited	Limited	Limited
Approval Change On Overrun	No	No	Yes, for labor plus material/ contract costs in overrun
Approval Status Change Definition	Yes	Yes	Yes
Cost Center Security	No	No	Yes
Contract Tracking	Yes, through the work order system showing breakdown of labor hours at zero dollar/hour and material used, or through the PO system showing only total dollars spent on contract maintenance	Yes, through the work order system showing break down of labor hours at zero dollar/hour and material used, or through the PO system showing only total dollars spent on contract maintenance	Yes, through a separate comprehensive module that accounts properly for contract PO and retains a separate history of labor hours at true cost/hour and material used by contractors at true dollar cost
Third-Party Billing, Invoice	No	Yes	Yes
Third-Party Billing, Mark Up L&M	No	Yes	Yes
Third-Party Billing, Mark Up Other	No	Yes	Yes
Third-Party Billing, Rounding	No	No	Yes
Third-Party Billing, Minimum Time	No	No	Yes

Question Keywords			
Company	Ashcom	Champs	Datastream Systems
Third-Party Billing, Minimum Charge	No	No	Yes
Third-Party Billing, Security	No	Configurable	Yes
Warranty Work Orders	Yes	Configurable	Yes
Warranty Claims	No	No	Yes
Multiple Warranties	Yes	Yes	Yes
Warranty Types	Yes	Yes	Yes
Calendar Warranties	No	No	Yes
Meter Warranties	No	Yes	Yes
Warranty Renewals	Yes	No	Yes
Serial-Based Warranty Expiration	No	Yes	Configurable
Nonserial-Based Warranty Expiration	No	Yes	Configurable

Work order control

Question Keywords			
Company	IFS North America	Indus International	Ivara
Resource Scheduling	Comprehensive	Comprehensive	Comprehensive
Resource Analysis	Some	Some	Some
Resource Costing	Some	Comprehensive	Some
Part Source Identification	Comprehensive	Comprehensive	Comprehensive
Approval Change On Overrun	No	No	No
Approval Status Change Definition	Yes	Yes	No
Cost Center Security	Yes	No	No
Contract Tracking	Yes, through the purchase order system showing only total dollars spent on contract maintenance	Yes, through a separate comprehensive module that accounts properly for contract PO and retains a separate history of labor hours at true cost/hour and material used by contractors at true dollar cost	Yes, through the purchase order system showing only total dollars spent on contract maintenance
Third-Party Billing, Invoice	Yes	Yes	No
Third-Party Billing, Mark Up L&M	Yes	Yes	No
Third-Party Billing, Mark Up Other	Yes	Yes	No
Third-Party Billing, Rounding	No	No	No
Third-Party Billing, Minimum Time	Yes	No	No
Third-Party Billing, Minimum Charge	No	No	No

Question Keywords			
Company	IFS North America	Indus International	Ivara
Third-Party Billing, Security	Yes	Configurable	No
Warranty Work Orders	Yes	Yes	Yes
Warranty Claims	Yes	Configurable	Configurable
Multiple Warranties	Yes	No	Yes
Warranty Types	Yes	Configurable	Configurable
Calendar Warranties	Configurable	No	Configurable
Meter Warranties	Yes	No	Yes
Warranty Renewals	Yes	No	Configurable
Serial-Based Warranty Expiration	No	Configurable	Configurable
Nonserial-Based Warranty Expiration	No	No	Configurable

Work order control

Question Keywords			
Company	MicroWest Software	MRO Software	SAP
Resource Scheduling	Comprehensive	Comprehensive	Some
Resource Analysis	Some	Some	Some
Resource Costing	Some	Comprehensive	Comprehensive
Part Source Identification	Limited	Comprehensive	Comprehensive
Approval Change on overrun	No	Yes, for labor plus material/contract costs in overrun	Yes, for labor plus material/contract costs in overrun
Approval Status Change Definition	No	Yes	Yes
Cost Center Security	Yes	Yes	Yes
Contract Tracking	Yes, through the purchase order system showing only total dollars spent on contract maintenance	Yes, through a separate comprehensive module that accounts properly for contract PO and retains a separate history of labor hours at true cost/hour and material used by contractors at true dollar cost	Yes, through a separate comprehensive module that accounts properly for contract PO and retains a separate history of labor hours at true cost/hour and material used by contractors at true dollar cost
Third-Party Billing, Invoice	Yes	Configurable	Yes
Third-Party Billing, Mark Up L&M	Yes	Configurable	Yes
Third-Party Billing, Mark Up Other	Yes	Configurable	Yes
Third-Party Billing, Rounding	Yes	Configurable	Yes
Third-Party Billing, Minimum Time	Yes	Configurable	Yes

Question Keywords			
Company	MicroWest Software	MRO Software	SAP
Third-Party Billing, Minimum Charge	Yes	Configurable	Yes
Third-Party Billing, Security	Yes	Configurable	Yes
Warranty Work Orders	Yes	Yes	Yes
Warranty Claims	Configurable	Configurable	Yes
Multiple Warranties	Configurable	Configurable	Yes
Warranty Types	Configurable	Configurable	Configurable
Calendar Warranties	Yes	Configurable	Configurable
Meter Warranties	Configurable	Yes	Yes
Warranty Renewals	Yes	Yes	Yes
Serial-Based Warranty Expiration	Configurable	Configurable	Configurable
Nonserial-Based Warranty Expiration	Configurable	Configurable	Configurable

Inventory control and purchasing

Question Keywords			
Company	Ashcom	Champs	Datastream Systems
LIFO Inventory	Yes	Yes	Yes
FIFO Inventory	Yes	Yes	Yes
Average Costing	Yes	Yes	Yes
Standard Costing	Yes	Yes	Yes
Activity-Based Costing	N/A	Simple ABC only	Third-party only
EOQ	Static field only	Single system-generated algorithms	Single system-generated algorithms
ABC Analysis	No	Yes	Yes
XYZ Analysis	No	Yes	Yes
Reorder Smoothing	No	No	No
Service Levels By Part	No	Yes	No
Service Levels By Category	No	No	No
What-If On Service Levels	No	Yes	No
Cycle Counts	Yes	Yes	Yes
Supplier History	Some aspects	Comprehensive	Comprehensive
Supplier Rating	Some aspects	Some aspects	Comprehensive
E-procurement	No	No	Comprehensive
History-Based Reorder	No	Yes	Yes
Lead Time-Based Reorder	No	Yes	No

Question Keywords			
Company	Ashcom	Champs	Datastream Systems
Max Level-Based Reorder	No	No	No
Component Tracking	No	Yes, all of the above tracking is possible	Yes, all of the above tracking is possible
New/Used Parts	Yes, some of this functionality is available through a workaround like using separate warehouses	No, you would have to create a separate part number for the used parts	Yes, but system does not favor used parts to be consumed first
Expected Delivery	Yes	Yes	Yes
Multiple Account Codes	No	Yes	Yes
Status Code	Yes	Yes	Yes
Multiple Ship-To	No	No	No
Tax Codes	Yes	Yes	Yes
Receipt Acknowledgement	Yes	No	Yes
Consignments	No	No	Simple

Inventory control and purchasing

Question Keywords			
Company	IFS North America	Indus International	Ivara
LIFO Inventory	Yes	No	No
FIFO Inventory	Yes	No	No
Average Costing	Yes	Yes	Yes
Standard Costing	Yes	No	No
Activity-Based Costing	Comprehensive	Simple ABC only	Third-party only
EOQ	Single system-generated algorithms	Multiple system-generated algorithms	Static field only
ABC Analysis	No	Yes	No
XYZ Analysis	No	Yes	No
Reorder Smoothing	No	Yes, smoothing algorithm is user-defined for different warehouse locations	No
Service Levels By Part	No	Yes	No
Service Levels By Category	No	Yes	No
What-If On Service Levels	No	Yes	No
Cycle Counts	Yes	Yes	Yes
Supplier History	Some aspects	Comprehensive	Some aspects
Supplier Rating	Comprehensive	Comprehensive	Some aspects
E-Procurement	Comprehensive	Comprehensive	No
History-Based Reorder	Yes	Yes	No
Lead Time-Based Reorder	Yes	Yes	No

Question Keywords			
Company	IFS North America	Indus International	Ivara
Max Level-Based Reorder	Yes	Yes	No
Component Tracking	Yes, all of the above tracking but not costing as described	Yes, all of the above tracking is possible	Yes, all of the above tracking but not costing as described
New/Used Parts	Yes, some of this functionality is available through a workaround like using separate warehouses	Yes, but system does not favor used parts to be consumed first	Yes, and system favors used parts to be consumed first
Expected Delivery	Yes	Yes	Yes
Multiple Account Codes	Yes	Yes	Yes
Status Code	Yes	Yes	Yes
Multiple Ship-To	Yes	Yes	No
Tax Codes	Yes	Yes	Yes
Receipt Acknowledgement	Yes	Yes	Yes
Consignments	Simple	Comprehensive	Simple

Inventory control and purchasing

Question Keywords			
Company	MicroWest Software	MRO Software	SAP
LIFO Inventory	No	No	Yes
FIFO Inventory	No	No	Yes
Average Costing	Yes	Yes	Yes
Standard Costing	Yes	Yes	Yes
Activity-Based Costing	Simple ABC only	N/A	Comprehensive
EOQ	Static field only	Single system-generated algorithms	Multiple system-generated algorithms
ABC Analysis	Yes	Yes	Yes
XYZ Analysis	Yes	No	Yes
Reorder Smoothing	No	No	Yes, smoothing algorithm is system-defined
Service Levels By Part	No	No	Yes
Service Levels By Category	No	No	No
What-If On Service Levels	No	No	No
Cycle Counts	No	Yes	Yes
Supplier History	Some aspects	Some aspects	Comprehensive
Supplier Rating	Some aspects	Some aspects	Comprehensive
E-procurement	Some	Comprehensive	Comprehensive
History-Based Reorder	Yes	No	Yes
Lead Time-Based Reorder	No	Yes	Yes
Max Level-Based Reorder		No	Yes

Question Keywords			
Company	MicroWest Software	MRO Software	SAP
Component Tracking	Yes, all of the above tracking but not costing as described	Yes, all of the above tracking is possible	Yes, all of the above tracking is possible
New/Used Parts	Yes, some of this functionality is available through a workaround like using separate warehouses	Yes, but system does not favor used parts to be consumed first	Yes, but system does not favor used parts to be consumed first
Expected Delivery	Yes	Yes	No
Multiple Account Codes	No	Yes	Yes
Status Code	No	No	Yes
Multiple Ship-To	No	No	No
Tax Codes	Yes	Yes	No
Receipt Acknowledgment	No	Yes	Yes
Consignments	Simple	Simple	Comprehensive

Preventive and condition-based maintenance I

Question Keywords			
Company	Ashcom	Champs	Datastream Systems
Standard PM Tasks	No	No	Yes (third-party)
Standard PM Task Times	No	No	Yes (third-party)
Standard Corrective Tasks	No	No	Yes (third-party)
Standard RCM Conditions	No	No	Yes (third-party)
Standard Safety	No	No	Yes (third-party)
PM Triggers	Yes, by all three	Yes, by all three	Yes, by all three
Multiple Condition Triggers	No	Yes	Yes
Automatic Trigger Reset	No	Yes	Yes
Nested Triggers	No	No	No
Combined Indicators	No	Yes	Yes
Recommend Corrective Action	No	No	Yes
Preferred Date	No	Yes	Yes
Forecast Rounds Schedule	No	No	Yes
PM Shadowing	No	Yes	Yes
PM Override	Yes	No	Yes
Reading Validation	No	Yes	Yes
Color-Coded Alarms	No	No	Yes (third-party)
Component/Indicator Hierarchy	No	No	Yes (third-party)

Question Keywords			
Company	Ashcom	Champs	Datastream Systems
Indicator Drill-Down	No	No	No
Graphic Highlights	No	No	Yes (third-party)
Alarm Acknowledge	No	No	Yes (third-party)
New Condition Entry	No	No	Yes (third-party)
History-Based PM	No	Yes, can generate reports manually and adjust manually as required	Yes, system automatically analyzes history and suggests corrections
Seasonal PM	No	Limited	Limited
Shutdowns/ Peaks/Holidays	No	No	Some

Preventive and condition-based maintenance I

Question Keywords			
Company	IFS North America	Indus International	Ivara
Standard PM Tasks	No	No	Yes
Standard PM Task Times	No	No	Yes
Standard Corrective Tasks	No	No	Yes
Standard RCM Conditions	No	No	Yes
Standard Safety	No	No	Yes
PM Triggers	Yes, by all three	Yes, by all three	Yes, by all three
Multiple Condition Triggers	Yes	Yes	Yes
Automatic Trigger Reset	No	Yes	Yes
Nested Triggers	No	Yes	Yes
Combined Indicators	No	No	Yes
Recommend Corrective Action	No	Yes	Yes
Preferred Date	No	No	Yes
Forecast Rounds Schedule	No	No	Yes
PM Shadowing	Yes	Yes	Yes
PM Override	Yes	Yes	Yes
Reading Validation	No	No	Yes
Color-Coded Alarms	Yes (third-party)	No	Yes
Component/ Indicator Hierarchy	Yes (third-party)	No	Yes

Question Keywords			
Company	IFS North America	Indus International	Ivara
Indicator Drill-Down	Yes (third-party)	No	Yes
Graphic Highlights	Yes (third-party)	No	Yes
Alarm Acknowledge	Yes (third-party)	No	Yes
New Condition Entry	Yes (third-party)	No	Yes
History-Based PM	Yes, can generate reports manually and adjust manually as required	Yes, can generate reports manually and adjust manually as required	Yes, can generate reports manually and adjust manually as required
Seasonal PM	Limited	Limited	Limited
Shutdowns/ Peaks/Holidays	All	No	Some

Preventive and condition-based maintenance I

Question Keywords			
Company	MicroWest Software	MRO Software	SAP
Standard PM Tasks	No	No	Yes (third-party)
Standard PM Task Times	No	No	Yes (third-party)
Standard Corrective Tasks	No	No	No
Standard RCM Conditions	No	No	Yes (third-party)
Standard Safety	No	No	Yes (third-party)
PM Triggers	Yes, by all three	Yes, by all three	Yes, by all three
Multiple Condition Triggers	Yes	Yes	Yes
Automatic Trigger Reset	Yes	Yes	Yes
Nested Triggers	No	No	Yes
Combined Indicators	No	No	Yes (third-party)
Recommend Corrective Action	Yes (third-party)	Yes	Yes (third-party)
Preferred Date	No	No	Yes
Forecast Rounds Schedule	Yes	Yes	No
PM Shadowing	Yes	Yes	Yes
PM Override	No	No	Yes
Reading Validation	No	Yes	Yes
Color-Coded Alarms	No	Yes	Yes (third-party)
Component/ Indicator Hierarchy	No	No	No

Question Keywords			
Company	MicroWest Software	MRO Software	SAP
Indicator Drill-Down	No	No	Yes (third-party)
Graphic Highlights	No	No	Yes (third-party)
Alarm Acknowledge	No	No	Yes (third-party)
New Condition Entry	No	No	Yes (third-party)
History-Based PM	Yes, can generate reports manually and adjust manually as required	Yes, can generate reports manually and adjust manually as required	Yes, can generate reports manually and adjust manually as required
Seasonal PM	Comprehensive	Limited	Limited
Shutdowns/ Peaks/Holidays	Some	Some	Some

Preventive and condition-based maintenance II

Question Keywords			
Company	Ashcom	Champs	Datastream Systems
PM By The Hour	No	Yes	No
PM By Recurrences	No	No	Yes
PM By Calendar	Yes	Yes	Yes
PM By Date	No	No	Yes
PM By Exception	No	No	Yes
PM Route By Asset Type	No	Yes	Yes
PM Route By Location	No	Yes	Yes
GIS Location	No	No	Some
Problem/Cause/Action	No	Yes	Yes
L&M By Asset	No	No	Yes
L&M Averaged	No	Yes	Yes
Critical Percent	No	No	No
Critical Route	No	No	No
Auto Priority Increase	No	No	No
Multiple Inspections	No	No	Yes
Risk/Criticality Record	Yes	No	Yes
Automatic Frequency	No	No	Yes
Confidence Rating	No	No	Yes
Expected Value	No	No	Yes
Historical Information	No	No	Yes
Track Inspection Tools	No	No	Yes
Critical Value Formula	No	No	Yes
WO/PM On Critical Value	No	No	Yes
Min/Max Values	No	Yes	Yes
Tolerance	No	Yes	Yes
Regression Analysis	No	No	Yes
Extreme Value	No	Yes	Yes

Preventive and condition-based maintenance II

Question Keywords			
Company	IFS North America	Indus International	Ivara
PM By The Hour	No	No	No
PM By Recurrences	Yes	Yes	Yes
PM By Calendar	Yes	Yes	No
PM By Date	No	Yes	No
PM By Exception	No	No	Yes
PM Route By Asset Type	No	Yes	Yes
PM Route By Location	Yes	Yes	Yes
GIS Location	Yes (third-party)	Yes (third-party)	Yes (third-party)
Problem/Cause/Action	Yes	No	Yes
L&M By Asset	No	No	Yes
L&M Averaged	No	Yes	No
Critical Percent	Yes	No	No
Critical Route	No	No	No
Auto Priority Increase	No	No	No
Multiple Inspections	Yes	Yes	Yes
Risk/Criticality Record	No	Yes	Yes
Automatic Frequency	No	No	No
Confidence Rating	No	No	No
Expected Value	No	Yes	Yes
Historical Information	Yes	Yes (third-party)	Yes
Track Inspection Tools	Yes	Yes	No
Critical Value Formula	No	No	Yes
WO/PM On Critical Value	Yes	Yes	Yes
Min/Max Values	Yes	Yes	Yes
Tolerance	No	No	Yes
Regression Analysis	No	No	Yes (third-party)
Extreme Value	Yes	Yes	No

Preventive and condition-based maintenance II

Question Keywords			
Company	MicroWest Software	MRO Software	SAP
PM By The Hour	No	No	Yes
PM By Recurrences	No	Yes	Yes
PM By Calendar	Yes	Yes	Yes
PM By Date	Yes	Yes	No
PM By Exception	No	No	No
PM Route By Asset Type	Yes	Yes	Yes
PM Route By Location	Yes	Yes	Yes
GIS Location	Some	Yes (third-party)	Yes (third-party)
Problem/Cause/Action	No	Yes	Yes
L&M By Asset	No	Yes	Yes
L&M Averaged	No	No	No
Critical Percent	No	No	No
Critical Route	No	No	No
Auto Priority Increase	No	Yes	No
Multiple Inspections	Yes	Yes	Yes
Risk/Criticality Record	No	No	Yes (third-party)
Automatic Frequency	No	No	Yes (third-party)
Confidence Rating	No	No	Yes (third-party)
Expected Value	Yes	Yes	Yes
Historical Information	No	Yes	Yes
Track Inspection Tools	Yes	Yes	Yes
Critical Value Formula	Yes (third-party)	No	Yes (third-party)
WO/PM On Critical Value	Yes (third-party)	Yes	Yes (third-party)
Min/Max Values	Yes	Yes	Yes
Tolerance	Yes	Yes	Yes
Regression Analysis	No	No	Yes (third-party)
Extreme Value	No	Yes	Yes

Equipment history

Question Keywords			
Company	Ashcom	Champs	Datastream Systems
Downtime Vs. Breakdown	Yes (with configuration)	No	Yes (with configuration)
Actual Vs. Budget	Yes (detailed)	Yes (detailed)	Yes (detailed)
Budget Integration	No	No	Yes
Budget What-If	No	Some	Some
User-Defined Costs	No	No	Yes
Problem/Cause/Action	No	Yes, but only certain problem codes are shown for a given component/ equipment	Yes, as above but also relevant cause codes are displayed for the selected problem code
Complaint Analysis	Third-party	No	Configurable
Root Cause Analysis	Third-party	No	Configurable
Remote db Synch	Third-party	No	Configurable
Browser Look and Feel	No	No	Configurable
Voice	Third-party	No	Configurable
Format Error-Check	No	No	Configurable
Range Error-Check	No	No	Configurable
Error Message Quality	No	No	Configurable
Custom Error Messages	No	No	Configurable
Report MTBF	Yes	Yes	Yes
MTBF by Root Cause	Third-party	Yes	Yes
Report MTTR	No	Yes	Configurable

Question Keywords			
Company	Ashcom	Champs	Datastream Systems
Report MWT	Third-party	No	Configurable
TSLF By Problem	No	No	Configurable
TSLF By Cause	Third-party	No	Configurable
TSLF By Action	No	No	Configurable
TSLF By Checkpoint	No	No	Configurable
Life Cycle Costing	Third-party	Yes	Configurable
Troubleshooting Database	No	No	Configurable

Equipment history

	Question Keywords		
Company	IFS North America	Indus International	Ivara
Downtime Vs. Breakdown	Yes (with configuration)	Yes	Yes
Actual Vs. Budget	Yes (detailed)	Yes (macro level only)	Yes (detailed)
Budget Integration	Yes	No	No
Budget What-If	No	No	No
User-Defined Costs	Yes	Yes	Yes
Problem/Cause/ Action	Yes, but only certain problem codes are shown for a given component/ equipment	Yes, as above but also relevant action codes are shown pertaining to the selected cause code, i.e., full nesting	Yes, as above but also relevant action codes are shown pertaining to the selected cause code, i.e., full nesting
Complaint Analysis	Configurable	Configurable	Configurable
Root Cause Analysis	Configurable	Configurable	Configurable
Remote db Synch	Configurable	Configurable	Configurable
Browser Look and Feel	No	Configurable	Configurable
Voice	Configurable	Configurable	Configurable
Format Error-Check	No	Configurable	Configurable
Range Error-Check	No	Configurable	Configurable
Error Message Quality	No	Configurable	Configurable
Custom Error Messages	No	Configurable	Configurable
Report MTBF	Yes	Yes	Configurable
MTBF By Root Cause	Yes	Yes	Configurable
Report MTTR	Yes	Yes	Configurable

	Question Keywords		
Company	IFS North America	Indus International	Ivara
Report MWT	Yes	No	Configurable
TSLF By Problem	Configurable	Configurable	Configurable
TSLF By Cause	Configurable	Configurable	Configurable
TSLF By Action	Configurable	Configurable	Configurable
TSLF By Checkpoint	Configurable	Configurable	Configurable
Life Cycle Costing	Yes	Yes	Yes
Troubleshooting Database	No	Third-party	Configurable

Equipment history

Question Keywords			
Company	MicroWest Software	MRO Software	SAP
Downtime Vs. Breakdown	Yes (with configuration)	Yes	Yes (with configuration)
Actual Vs. Budget	Yes (detailed)	Yes (detailed)	Yes (detailed)
Budget Integration	No	No	Yes
Budget What-If	No	No	Some
User-Defined Costs	No	Yes	Yes
Problem/Cause/ Action	Yes, but only certain problem codes are shown for a given component/ equipment	Yes, as above but also relevant action codes are shown pertaining to the selected cause code, i.e., full nesting	Yes, but only certain problem codes are shown for a given component/ equipment
Complaint Analysis	Configurable	Configurable	Third-party
Root Cause Analysis	Configurable	Configurable	Third-party
Remote db Synch	Configurable	Configurable	Third-party
Browser Look and Feel	Configurable	Configurable	Third-party
Voice	Yes	Configurable	Third-party
Format Error-Check	Configurable	Configurable	Configurable
Range Error-Check	Configurable	Configurable	Configurable
Error Message Quality	Configurable	Configurable	Configurable
Custom Error Messages	Configurable	Configurable	Third-party
Report MTBF	Yes	Yes	Yes
MTBF By Root Cause	Configurable	Yes	Configurable
Report MTTR	No	Configurable	Yes

Question Keywords			
Company	MicroWest Software	MRO Software	SAP
Report MWT	No	Configurable	Configurable
TSLF By Problem	No	Configurable	Configurable
TSLF By Cause	No	Configurable	Configurable
TSLF By Action	No	Configurable	Configurable
TSLF By Checkpoint	No	Configurable	Configurable
Life Cycle Costing	Yes	Configurable	Yes
Troubleshooting Database	Configurable	Yes	No

Advanced functions and displays I

Question Keywords			
Company	Ashcom	Champs	Datastream Systems
Trigger On PM	No	Yes	Configurable
Trigger On Limits	No	No	Configurable
Trigger On Downtime	No	No	Configurable
Trigger On MTBF	No	No	Configurable
Trigger On Warranty	No	No	Configurable
Trigger On Parts Missing	No	No	Configurable
Trigger On Wait Time	No	No	Configurable
Trigger On Activity Times	No	No	Configurable
Trigger On PM Critical %	No	No	Configurable
Trigger On Expenditure	No	No	Configurable
Trigger On Repeat Failure	No	No	Configurable
Lookup Equipment	No	Yes	Yes
Lookup Parts	No	No	No
Lookup Suppliers	No	No	No
Lookup Employees	No	No	No
Lookup Work Orders	No	No	No
Lookup Online Help	Yes	No	Yes
Lookup Corporate Structure	No	No	No
Lookup Warehouses/Stores	No	No	No
Lookup Projects	No	No	No
Lookup KPIs	No	No	No
Lookup Menus/Functions	Yes	Yes	Yes
Lookup G/L Accounts	No	No	No
Toggle	N/A	Comprehensive	N/A

Advanced functions and displays I

Question Keywords			
Company	IFS North America	Indus International	Ivara
Trigger On PM	Yes	Yes	Yes
Trigger On Limits	No	No	Yes
Trigger On Downtime	Yes	No	Yes
Trigger On MTBF	No	No	Configurable
Trigger On Warranty	Yes	Yes	Configurable
Trigger On Parts Missing	Yes	Yes	Configurable
Trigger On Wait Time	Yes	No	No
Trigger On Activity Times	Yes	No	No
Trigger On PM Critical %	Yes	No	Configurable
Trigger On Expenditure	No	No	Configurable
Trigger On Repeat Failure	Yes	No	Configurable
Lookup Equipment	Yes	No	Yes
Lookup Parts	Yes	No	Yes
Lookup Suppliers	No	No	No
Lookup Employees	Yes	No	Yes
Lookup Work Orders	Yes	No	No
Lookup Online Help	No	No	Yes
Lookup Corporate Structure	No	No	Yes
Lookup Warehouses/Stores	No	No	No
Lookup Projects	Yes	No	Yes
Lookup KPIs	No	No	Yes
Lookup Menus/Functions	Yes	No	No
Lookup G/L Accounts	No	No	No
Toggle	Equipment/ parts only	N/A	Equipment/parts only

Advanced functions and displays I

Question Keywords			
Company	MicroWest Software	MRO Software	SAP
Trigger On PM	No	Configurable	Yes
Trigger On limits	No	Configurable	Configurable
Trigger On downtime	No	Configurable	Yes
Trigger On MTBF	No	Configurable	Yes
Trigger On warranty	No	Configurable	Configurable
Trigger On Parts Missing	No	Configurable	Yes
Trigger On Wait Time	Yes	Configurable	Yes
Trigger On Activity Times	No	Configurable	Configurable
Trigger On PM Critical %	No	Configurable	No
Trigger On Expenditure	No	No	Yes
Trigger On Repeat Failure	No	Configurable	Configurable
Lookup Equipment	Yes	Yes	Yes
Lookup Parts	Yes	Yes	Yes
Lookup Suppliers	Yes	No	No
Lookup Employees	Yes	No	No
Lookup Work Orders	No	No	Yes
Lookup Online Help	Yes	Yes	No
Lookup Corporate Structure	No	No	Yes
Lookup Warehouses/Stores	Yes	Yes	No
Lookup Projects	No	No	Yes
Lookup KPIs	No	No	No
Lookup Menus/Functions	No	No	Yes
Lookup G/L Accounts	No	No	Yes
Toggle	N/A	Equipment/ parts only	Equipment/ parts only

Advanced functions and displays II

Question Keywords			
Company	Ashcom	Champs	Datastream Systems
Business Intelligence	Third-party	No	Comprehensive
Balanced Scorecard	No	No	Comprehensive
KPI Reports	No	No	Comprehensive
PDA WO Integration	Yes	Yes	Yes
PDA Work	Yes	Yes	Yes
PDA Parts Required	Yes	No	Yes
PDA Inventory	Yes	No	Yes
PDA Parts Used	Yes	Yes	Yes
PDA Tools Required	Yes	No	Yes
PDA Scanner	Yes	No	Yes
PDA Work Requests	No	No	Yes
Fleet VMRS	Yes	Yes	Yes
Fleet Fuel	Yes	Yes	Third-party
Vehicle Mileage	Yes	No	Yes
Vehicle Utilization	Yes	No	Yes
Vehicle Status	No	No	Yes
Vehicle Environment	No	No	Third-party
Vehicle Wear Analysis	Yes	No	Yes
Vehicle Reservations	Yes	Yes	Yes

Question Keywords			
Company	Ashcom	Champs	Datastream Systems
Fleet Depreciation	Yes	Yes	Yes
Vehicle Operation	Yes	Yes	Yes
IT Asset Management	No, not an area of specialization for the CMMS	No, not an area of specialization for this CMMS	Yes, some of the technology asset management features above
Capital Asset Planning	No, not an area of specialization for the CMMS	Yes, some of the capital asset planning features above	Yes, some of the capital asset planning features above
Tool Management	Yes	No	Yes
Key Management	Yes	No	Configurable
Lockout/Tagout	No	Yes	Configurable
Other Function/Status	Web work request, DocuMizer (OCR-Based data entry)		Calibration

Advanced functions and displays II

Question Keywords			
Company	IFS North America	Indus International	Ivara
Business Intelligence	Comprehensive	Some	Some
Balanced Scorecard	Comprehensive	Some	Some
KPI Reports	Comprehensive	Comprehensive	Comprehensive
PDA WO Integration	Yes	Configurable	Yes
PDA Work	Yes	Configurable	Yes
PDA Parts Required	Yes	Configurable	Yes
PDA Inventory	Yes	Configurable	Configurable
PDA Parts Used	Yes	Configurable	Yes
PDA Tools Required	Yes	Configurable	Yes
PDA Scanner	Yes	Configurable	Yes
PDA Work Requests	Yes	Configurable	Yes
Fleet VMRS	No	No	No
Fleet Fuel	Yes	Yes	Yes
Vehicle Mileage	Yes	Yes	Yes
Vehicle Utilization	Yes	No	No
Vehicle Status	No	No	Third-party
Vehicle Environment	No	Third-party	No
Vehicle Wear Analysis	No	Yes	Yes
Vehicle Reservations	Yes	Yes	No

Question Keywords			
Company	IFS North America	Indus International	Ivara
Fleet Depreciation	Yes	Third-party	No
Vehicle Operation	Yes	Yes	Yes
IT Asset Management	Yes, some of the technology asset management features above	Yes, some of the technology asset management features above	Yes, some of the technology asset management features above
Capital Asset Planning	Yes, some of the capital asset planning features above	No, not an area of specialization for the CMMS	Yes, some of the capital asset planning features above
Tool Management	Yes	Yes	Yes
Key Management	No	Configurable	Configurable
Lockout/Tagout	Configurable	Yes	Configurable
Other Function/ Status		Personnel Qualification, Permit Tracking, Document Management, MSDS, Procurement Eng., Engineering Change	Asset prioritization, RCM2 implementation

Advanced functions and displays II

Question Keywords			
Company	MicroWest Software	MRO Software	SAP
Business Intelligence	Configurable	Comprehensive	Comprehensive
Balanced Scorecard	Configurable	No	Some
KPI Reports	No	Comprehensive	Configurable
PDA WO Integration	Yes	Yes	Yes
PDA Work	Yes	Yes	Yes
PDA Parts Required	Yes	Yes	Yes
PDA Inventory	Configurable	Yes	Yes
PDA Parts Used	Yes	Yes	Yes
PDA Tools Required	Yes	Yes	No
PDA Scanner	Yes	Yes	Yes
PDA Work Requests	Yes	Yes	Yes
Fleet VMRS	Third-party	Yes	Third-party
Fleet Fuel	Yes	Yes	Yes
Vehicle Mileage	Yes	Yes	Yes
Vehicle Utilization	No	Yes	Yes
Vehicle Status	No	Yes	Third-party
Vehicle Environment	No	No	No
Vehicle Wear Analysis	Yes	Yes	Yes
Vehicle Reservations	No	Yes	No

Question Keywords			
Company	MicroWest Software	MRO Software	SAP
Fleet Depreciation	No	Yes	Yes
Vehicle Operation	Yes	Yes	Yes
IT Asset Management	Yes, some of the technology asset management features above	Yes, a separate module auto-polls IT assets for maintaining s/w and h/w demographics database, device and license tracking, depreciation mgmt, lease and contract mgmt and repair mgmt, utilization reporting, help-desk mgmt, etc.	Yes, some of the technology asset management features above
Capital Asset Planning	No, not an area of specialization for the CMMS	No, not an area of specialization for the CMMS	Yes, some of the capital asset planning features above
Tool Management	Yes	Yes	Configurable
Key Management	Yes	Configurable	No
Lockout/Tagout	Configurable	Yes	Yes
Other Function/ Status		Maximo Industry Solutions including Pharmaceutical, Calibration, Oil & Gas, Transmission & Distribution, Nuclear, and Transportation	

SOURCE: Plant Services, www.plantservices.com/cmms_review

Chapter

5

How to Implement a CMMS

Introduction

A well-planned and executed *computerized maintenance management system* (CMMS) project can yield a maximum *return on investment* (ROI) realized through increased efficiency, productivity, and profits. However, a poorly planned and executed CMMS project can result in a loss of revenues. These losses can be measured in terms of the overall investment in the project, as well as from wasted time, and lost projected revenue forecast upon the successful installation and implementation of a CMMS. This chapter outlines how to successfully implement a CMMS.

Why So Many CMMS Projects Fail

Many CMMS projects fail to reach their full potential and many just plain fail. Here are some of the factors:

- *Not having management support for the CMMS.* The major element necessary to the success of any large undertaking is commitment to the project and support by upper-level management. Lack of interest on the part of upper-level management will diminish the chances of success. If upper-level management approaches it from a rational, reasonable perspective, and provides necessary resources, success is almost assured.
- *Employee turnover.* CMMS projects fail because of employee turnover for one of the three reasons: A key member, or members of a project team may resign, be terminated, or transferred. With a sophisticated project like the CMMS, continuity is a key factor in its success. In order to establish that continuity, and maintain it in the event of personnel

changes, each step of the project should be fully and accurately documented.

- *Employee resistance.* Often, employee resistance to computers is not considered when management decides to acquire a CMMS. This problem can be more devastating than losing key members of the project team. Employees may accept the computerization enthusiastically or become hostile to the idea. Management may look at the CMMS as a tool to help employees in their work, and in turn enhance the bottomline. Employees, on the other hand, may view the CMMS as an intrusion, threatening their professional and personal security.

  Users need to understand that there will be a transition process, the current process will change and at least in the short term it may be painful. CMMS projects that promise efficiency improvements are often looked upon suspiciously by technicians as excuses for downsizing.

  CMMS projects fail not necessarily because of the system, but rather as a result of refusal by the users to use it. Ultimately, success will be determined by the amount of system usage. Not using the system, obviously leads to failure.

  Employee resistance to the computerization does not have to prevent or delay your project. Thoroughly plan your implementation phase. Be familiar with the new system. Actively lead your employees. Make it clear you would not ask them to do something you would not do yourself. Reinforce the importance of their support and the overall commitment to the project. Emphasize that the CMMS will be a helpful tool and should not be considered a threat either personally or professionally. Make sure your employees know they have a receptive place to turn if there are problems. If you take these steps, the chances of your system living up to or passing your expectations are much higher.

- *Wrong selection of the CMMS.* This is one of the top reasons of failures. There is nothing like a best CMMS package. You have to select the right CMMS based on your needs and requirements. A CMMS designed for manufacturing environment may not work for a healthcare facility. Chapter 4 discusses this in great depth.

  When companies started the process of acquiring a CMMS back in early to mid 80s, they were not as educated about CMMS and the technology as they are today. As a result, lot of organizations ended up with the wrong package for their application. Do not feel bad, as there is always a bright side to every thing. In the process, organizations have become more educated regarding CMMS and now they know exactly what they want or what they do not want. So, when they are ready for an upgrade to CMMS, hopefully, they will make a right selection.

- *Justifying based on advanced functionalities.* Implementation typically has two phases of progression—primary and advanced functionalities. Most companies achieve the primary functionalities phase and very few reach the advanced phase. Unfortunately, most organizations justify their CMMS solution based on the advanced phase. You should try to justify your CMMS based on achieving primary functionalities. Further achievements are a bonus that will make the overall implementation of your CMMS a great success.
- *Being locked into restrictive hardware/software.* Sometimes corporate policies dictate hardware as well software requirements. The best CMMS you find may not work on company required hardware, or a particular CMMS is required to be used by all their facilities. Some of these policies make sense but some times do not work well as needs of each facility might be different, requiring different solutions.
- *Lack of adequate training during implementation.* If users do not know how to use the software effectively, you will not have a successful implementation. Training of users is very important. You will see more details regarding training further in the chapter.
- *Lack or absence of follow up and monitoring.* This goes back to lack of upper management commitment. Proper follow-up of the project to ensure the continuity is important.
- *Not having adequate supplier support for the CMMS.* This goes back to wrong selection. The best of CMMS will not work well if you do not get vendor support. That is one of the selection criteria.

It is interesting to review aspects that lead to successful implementations of a CMMS.

Table 5.1 shows a list of factors influencing CMMS implementation success.

As you can see the most important factors listed in their success were senior management commitment and effective training.

Step-By-Step Process for Implementing a CMMS Project

See Fig. 5.1.

Form a team

Establish a team. The first step in establishing a team is to select the team members. Many times the team has too many IT personnel. While having some IT representation is important, the team should primarily consist of representatives from the CMMS user community. Users bring

TABLE 5.1 Success Factors for CMMS Implementation

What do you consider are the two most important aspects of your implementation that led to your success?			
	Responses		
Factor	Most important	Second most important	Total
Senior management commitment	15	17	32
Effective training	12	17	29
Choosing the right CMMS	10	7	17
Effective change management	10	5	15
Focus on business benefits	5	9	14
Adequate budget	6	8	14
Effective BPR	5	8	13
Effective project management	5	5	10
CMMS Vendor support	7	2	9
Consultant support	4	2	6

SOURCE: plant-maintenance.com

to the table a better understanding of the process, needs, wants, and problems. This is important in order to gain end-user acceptance. It should be reinforced that IT is a facilitator of the solution (a service provider) and not the driver.

So, the team should consist of the plant engineer, maintenance manager, maintenance employees, and representatives from IT, operations, purchasing, and accounting departments. Marketing, sales, and human resources should also be included. You should get everyone involved who has any impact on or of this project. You might consider the software vendor on the team also. Software vendors will prove to be an asset on the team as they work with hundreds of companies.

Involving your employees in the implementation process enables you to break down their resistance to computers and build enthusiasm for CMMS as a tool to facilitate their work.

Systematic, periodic reports should be submitted to upper-level management keeping them informed of progress, or lack of it. If it is necessary to replace members of the project team, new team members can be introduced with a minimum of disruption to maintain the continuity of the project.

External resources. Considering outside consulting company to help implement the CMMS package is a viable option. An experienced implementation consultant can help guide you through the implementation by facilitating the decision-making process and helping you to avoid common pitfalls that tend to cause implementation failures. They typically bring years of experience and have insights that can be of significant value to the project.

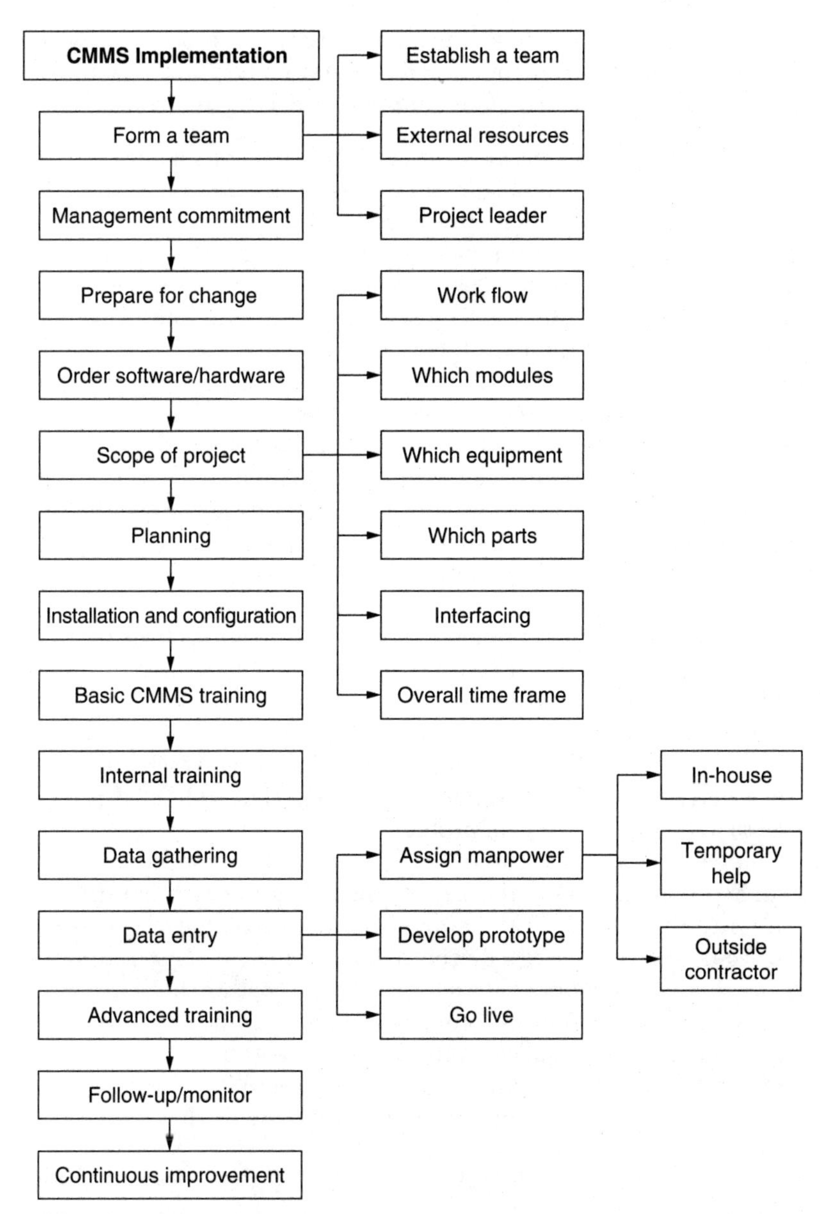

Figure 5.1 CMMS implementation/step-by-step process.

If you use a consultant, make sure all activities are properly documented and handed over to you. Avoid a situation where a consultant is gone and has left you with inadequate or no documentation.

Project leader. Whether you use your in-house team or an outside consultant, the maintenance manager should assign one person as the project leader responsible for implementing the CMMS.

Along with obvious commitment and enthusiasm for the new project, it is important that the project leader knows as much about the hardware and system as possible. The project leader should know the company structure, understand maintenance functionalities, and should be able to work with other departments. The team needs definite direction. Confidence in the project and its objectives come from a thorough knowledge of the application. If the leader can display such a knowledge and confidence to team, they will be motivated to learn as much as possible. The team should know what to expect from the system and what is expected from them. It should be understood that there would be problems in the beginning stages of the new system, but they will be worked out.

One of the project leader's roles is to keep all activities coordinated. As an example, an integrated CMMS will use inventory part numbers that will be used on work orders. If purchasing department decides to change its numbering system, all the part numbers and equipment parts list in the CMMS will be invalidated and useless. Work order and inventory activities will not function properly. The role of a project leader is to avoid these kinds of problems.

The other role is to keep the team motivated. One method of motivating and preparing the team for the change is to get them directly involved in the project from the start. Make it clear that changes are going to occur, and they have a direct role and input into those changes. If they feel that they not only have a stake in the program, but have some say in it as well, they will more likely take an active interest in the new system. Reward their initiative, and attempt to establish open lines of communication. Solicit ideas and feedback on a regular basis. Following are some points for encouraging communications with the team:

- Ask for suggestions concerning needs and desires for the new system.
- Allow your employees to help design custom input screens for the new system; this will give them some say in what they are going to be dealing with.
- Give your employees a forum to air their feelings and concerns about the new system. They are sure to have questions about their place in any reorganization changes that may occur as a result of the new system. Remember, communications is a two-way street. Listen carefully

to what they have to say. Consider their suggestions, concerns, and complaints.

- Periodically call your staff together to discuss training, progress, and any problem areas that may arise during the implementation stage and after.

Management commitment

As stated earlier, upper-level management must be totally committed to the CMMS project. This commitment must include allocation of manpower and resources needed to complete the project successfully. When providing for resources, it is extremely important that upper-level management understands that the purchase and operating costs of the project may sometimes be less than the cost of implementing it. It is crucial that you do not attempt to squeeze your manpower, or attempt to cut corners in the data-gathering and training phases of the project. If the data-gathering portion of your project requires more personnel, there should be contingency plans to provide the personnel in a timely fashion. Placing unrealistic expectations on your project team will only prove counter productive. Although the early stages of the project are time consuming and expensive, it is important to understand that patience and investment will pay off in the long run.

The role of upper-level management. During the implementation, plant engineers or middle managers will inevitably reach a hurdle that can only be cleared with the support of a senior manager. Other departments will have their own agendas and priorities and they will not see things your way. In these circumstances conflicts of interest are common. These disputes may only be resolved with the support of a senior manager.

Prepare for change

People do not change the way they work unless they understand the need for change and benefits to them in making the change. It is important they feel they are a part of setting the direction for change. You will be successful in bringing about changes to maintenance processes by involving key maintenance personnel in developing the new procedures.

Order software/hardware

During the justification process, you justified a CMMS and any hardware needed to support the project. Hardware includes PCs, servers, handheld devices, bar code and radio frequency identification device(RFID) readers, and accessories.

During the evaluation/selection process you selected a CMMS package suitable for your organization. Software includes application package (CMMS), bar code, and RFID. In some cases you may need operating system and database software (e.g., MS SQL and Oracle) and some other software to support the application. At this time you should issue purchase orders for software and any required hardware.

Define scope of project

You first review the existing workflow in detail and decide what changes you would like to make to improve productivity. Next, you need to determine which CMMS modules to implement. For example, you might implement equipment, PM, work order, and inventory modules first and purchasing module at a later date. In some cases, you might want to implement all of them simultaneously.

Then you should decide which equipment to implement first; all of them or the critical ones first to start with. Same concept applies to inventory parts. It basically depends upon return on investment and kind of manpower you have available.

At this point, you should also review opportunities to interface your CMMS to other systems such as purchasing or energy management. By interfacing to other systems, in most cases, you can further enhance productivity significantly. For example: Instead of filling out time cards manually, interface your CMMS to payroll software.

Once you have determined the scope of the project, determine the overall time frame to implement the project.

Planning

Properly planning the CMMS implementation project is one of the key elements. In the planning phase you determine the "what," "why," "who," and "how."

Equipment data

- *Determine equipment-numbering scheme.* Check if there is a scheme being used already. If the existing system is working fine, there is no need to change. If it is not, you need to come up with a numbering scheme. Think through carefully to include all different equipment you have in-house in your numbering scheme. As an example, if you are dealing with facilities, you could use XX-WF-000 for water filters where XX are alphabets for the building, WF stands for water filter, and 001 is numeric sequence. So, PH-WF-001 would be first water filter in physics building. Remember, at this stage, you are only planning, not actually gathering the data.

- *Determine equipment hierarchy.* Set up a parent/child relationship. For example, an air-handler can have pumps and motors as children. If you are going to keep track of parent and children both, you need to document the hierarchy. If you are dealing with a facility, a building can be parent, each floor can be a child of the building, and each room on the floor can be a child of that floor. Make sure to include every equipment/asset that falls under the hierarchy scheme.
- *Downtime monitoring.* Decide which equipment you are going to monitor. Then decide on criteria and how exactly are you going to do it. Make sure you have a way to differentiate between planned and unplanned downtime.
- *Spare parts.* Identify sources of information for spare parts pertinent to each piece of equipment.
- *Documents to be linked to equipment records.* Decide as to what kind of documents, drawings, and Web sites you want to link to each equipment. For example, each equipment record should be linked to manufacturer and vendor's Web sites.

Plan on collecting all information pertinent to impact of loss of operation or related data necessary to help you make decisions regarding both preventive and corrective maintenance actions. You should have this information for each piece of equipment or at least for critical equipment.

See App. 5A for a successful implementation case study.

Preventive maintenance. The following decisions have to be made for each PM task:

- Will the PM be performed by calendar time or run time (miles, hours, and so on)?
- Follow fixed schedule (regardless of completion date) or schedule based on completion date.
- How often PM work orders will be generated (daily, weekly, monthly, and so forth)?
- Route PM—this was explained in Chap. 2. It is important to decide strategies for route PMs under planning phase.

Procedures. Procedures can be preventive maintenance, safety instructions, or any other set of instructions. Each piece of equipment should have identified for it, all the preventive, corrective, and predictive maintenance tasks necessary to properly maintain that equipment. Along with the maintenance task, information regarding maintenance frequency, responsible craft, repair and/or consumable parts necessary to complete the maintenance task, and time estimated to complete the task are some of the additional information that will enhance the usefulness of your CMMS database. These procedures can then be used anywhere applicable within CMMS.

Labor. You need information on each maintenance technician such as name, address, phone, and social security. Review if the information is available as part of your existing system. If not, you might have to obtain it from HR or payroll department. You also have to decide if you are going to use some sort of ID card for your technicians that can be scanned by a bar-code reader or some other device. If yes, will the cards be produced in-house or by an outside vendor?

Inventory. The following decisions have to be made:

- *Determine part-numbering scheme.* Check if there is a scheme being used already. If the existing system is working fine, there is no need to change it. If it is not, you need to come up with a numbering scheme. Think through carefully to include all different parts you have in-house in your numbering scheme. There are companies who use 20 to 30 characters long part-numbering scheme. It includes every detail of that part. For example, type of part, thickness, diameter, and location. With advances in CMMS and a field available for each of these details (category, dimensions, location, and the like), you do not need a part-numbering scheme to include all the details. It just increases the data-entry error potential.
- *Determine location for each part.* Do you have a single stock room or multiple stock rooms? Are you going to continue the same way or planning to change. Within a stock room, do you have a location scheme that is working? Or do you need to design one? For example, aisle, bins and shelves.
- *Vendors (primary and secondary).* In data gathering phase you will compile a list of all the vendors you buy parts and services from. One of them has to be assigned as a primary vendor. CMMS automatically generates POs to the primary vendors, which can be changed if desired by the user. At this stage, you have to decide the criteria for primary vendors. Is it based on price, delivery, overall service, or combination of all?
- *Issue units.* Decide if you are going to use metric units (e.g., kilograms, meters, and liters) or British units (pounds, feet, and gallons), or a combination of both. How are you going to handle the inventory of pipes, beams, and the like. Are you going to keep track of lengths?

For example, if a pipe is cut, are you going to keep track of leftover pipe length?

- *Physical inventory process.* You have to make the following decisions:
  - How often are you going to take the physical inventory? Once or twice a year?
  - Who is going to do it?
  - Is it going to be manual or using mobile technology? If it is manual, make sure your CMMS prints appropriate forms for this purpose. If not, make sure mobile hardware and software is in place and the technicians have gone through proper training.
- *Documents to be linked to parts' records.* Decide as to what kind of documents, drawings and Web sites you want to link to each part. For example, each part record should be linked to manufacturer and vendor's Web sites.
- *Order parts.* Decisions have to be made on:
  - Who has the authority to order the parts?
  - Up to what amount?
  - Beyond what amount do you need further approval?
- *Cross-referencing requirements.* You need a database of company part numbers versus manufacturer/vendor part numbers. In planning phase, you need to decide the source of manufacturer and vendor part numbers.
- *Parts label design.* What information do you want to print on the parts labels? Part number, description, and location are typical.
- *Bar-code label design.* If you are using bar coding with inventory, you have to decide which information items you want bar coded—part number and location are typical.

If you have multiple plants/facilities, it is important every facility is following the same schemes. Without consistency, CMMS will not be very effective. If you are looking for a part at a different facility and if they are describing it differently than you are, you will not find the part even if it is in stock.

Purchasing and accounting. You will probably need purchasing and accounting departments involved for these:

- "Bill to" information
- "Ship to" information
- Sales tax rate
- Determine budget accounts and amounts

Codes. Plan and design codes to be used throughout CMMS. During planning phase, you need to decide on strategies as to basis of codes. Actual compilation of codes will be done during data gathering phase. Decide on the following:

- Account codes
- WO type
- Failure codes
- Action codes
- Repair codes
- WO priority
- Equipment criticality
- Work order status
- PO status
- Departments

Candidates for duplicate information. Many equipment are identical. The only difference being serial number. Many are similar. In either case, once you have one record created, others can be created by duplicating the first one and editing it. It saves you tremendous amount of data entry time. You should identify candidates for duplicate information, that is, identify all equipment, and inventory parts that are identical or similar.

Mobile applications. During the justifications and selection process, you identified mobile applications with your CMMS. Now is the time to take a closer look at those applications. Identify each application and work on details. Some examples are:

Equipment meter reading. You can use a hand-held remote data entry device to collect equipment meter readings.

Stock issue/return. Parts can be issued and/or returned using a hand-held device.

Receiving. Receiving allows stocked materials to be received using the hand-held device

Physical inventory cycle count. You can use hand-held devices to count and track inventory parts.

Backup. Following decisions have to be made:

- Back-up scheme (hardware and set up).
- How often backup will be done (daily, weekly, and so on)?
- Who will be assigned to do the job?

Archive/Merging. Answer these questions:

- Do we need to archive work order history file?
- When? For example, after the database has certain number of records (e.g., 15,000).
- When you archive, what kind of scheme? For example, archive the first six months of history; or you can decide to always keep last three years history in your active database. Make sure the archived database can be merged back.

History. Plan on what kind of maintenance history you want to maintain. Your decisions should include date performed, task performed, person performing, estimated and actual time to perform, equipment performed on, material used, and any outside contractor cost incurred.

Decisions. The following decisions have to be made:

- Are you going to print estimated time on work orders?
- Details of reports needed in your CMMS.
- Which reports are needed in graphics format?
- What decisions will be made based on reports analysis?
- Security issues: Who is permitted to do what?
- Screen layouts: Details of each screen and who is going to do it.
- Field labels: Do you need to change the terms provided by your CMMS on any of the label? If yes, document changes.

Analysis. CMMS can provide valuable information to assist with the following analyses. Make sure you use the CMMS to obtain these:

- The number of failures.
- The root causes of those failures.
- The maintenance costs associated with those failures.
- The production costs associated with those failures—note that these may incorporate more than just downtime costs.
- Any safety or environmental implications associated with those failures.

Key performance indicators (KPIs). During the justification process, you had developed a list of KPIs for your application. At this stage, you should review them and revise if necessary. You should also determine as to how you would compute those KPIs. Most of the KPIs should come from CMMS reporting.

Assign responsibilities. Plan on who will:

- Install the hardware (if necessary)
- Maintain the computer hardware, backups, and so forth
- Perform archiving and merging of data
- Take care of disaster recovery
- Generate reports
- Review and analyze various reports
- Plan and schedule work orders
- Do the ongoing data entry
- Close work orders
- Be responsible for customizing, configuring, tailoring, and maintaining the CMMS

Training

Training is a multiple-phase process.

Basic training. Make sure people who will operate the CMMS are familiar with computer basics and operating system. If not, provide them with basic training so they are ready for CMMS application.

Application (CMMS) training. CMMS application training can also be done in two different phases, initial training to get the system up and running, and advanced training after spending a period of time with CMMS. Timing of training should be kept in mind. Do not train too early. Training should be coordinated with implementation. Trainees should leave the class and use what they have learnt immediately.

Application training includes mobile technology, bar code, and RFID. (if your project includes these).

Internal training. You also need to train CMMS users with internal processes. This is often ignored in most CMMS implementation projects. For example, equipment numbering scheme: CMMS users should be trained on numbering schemes to avoid erroneous data entry. Another good example is inventory part description. If part description data entry is not consistent, you will find a corrupt inventory database after a while. Work order data entry is another example. Descriptions such as “machine down” or “ does not work” followed by a repair description of “done” or “fixed” will not give you meaningful history. Therefore, proper and consistent job description training is important.

General training guidelines. Be sure that every trainee is given enough help to become comfortable and confident in what he or she is doing. Even a small shortage of needed training can cause an employee to backslide, lose confidence, and eventually cause project failure. The new system cannot afford that. All personnel who will use the CMMS, or maintain it, or oversee its operation should be included in the training process.

Accept any and all training support from your vendor. You may wish to contract with your vendor for additional training support or follow-up training programs or services. Remember, while the training may initially seem an expensive proposition, in the long run it will not only prove beneficial, but very cost effective.

Training should be an on-going process to promise new users in your company with the full course while current users take refreshers as needed or desired. This continuity will accommodate the inevitable personnel changes and system evolutions that will occur over the life of the system. As users leave, their replacements must be trained as thoroughly as if they had been original users. This need is frequently overlooked!

During the training process, the newly installed CMMS can be checked thoroughly to assure that it works as planned. Feedback from trainees should be recorded and analyzed to assist in evaluating system performance and potential modification. Action on trainee responses will not only result in system refinements, but will improve the general acceptance of the CMMS.

Installation and configuration

You must make sure that necessary hardware and software other than CMMS is in place. For example, you might need new hardware or need to upgrade memory on your existing hardware. You also need to make sure of appropriate printers, and the like. Now, you need to install your CMMS. If it is client-server-based system, installation could be quite involved, as you have to install and configure the CMMS application on each client. If it is web-based, installation process is much simpler. If you have chosen a web-based *application service provider* (ASP) option, there is hardly any installation on your part.

Data gathering

Before gathering any data, check if you already have data in electronic format (say, if you are migrating from an existing system to a different system, you have the data in your computer). It may be possible to transfer this data electronically without retyping. Explore that possibility first. Your CMMS vendor and/or your IT department are the best source for that project. If you have corrupt data, you have two choices:

- Get your data cleansed if you can justify the cost (there are many companies that provide this kind of service, however, it is beyond the scope of this book).
- Revise the data (from printed outputs) and reenter the corrected data into the new system. Some "cutting and pasting" might be possible to save time.

If you do not have any existing data then you have to go through the process of collecting the data. Data collection and entry are critical to a successful implementation. The same care and attention given to planning the training effort must be given to locating data sources and determining how to collect needed data for entry into the CMMS.

You have to decide who is going to gather the data. Is it going to be an in-house team or an outside consultant? Once that is decided, you start the process of gathering data.

Equipment data. The first thing you need to do is make a list of all the equipment in-house. The current crews and supervisors will be able to contribute good knowledge of what equipment items are present in what areas, and accounting department records may help as well in developing the list.

Then you begin the data collection of equipment and preventive maintenance. Many sources of data may be used in the collection and entry process. The current crew of mechanics probably will have the best knowledge of what specific preventive maintenance work is needed to keep the equipment operating. Manufacturer specifications are valuable as well, and some CMMS suppliers offer their own databases of equipment performance history as both starting point and continuing reference.

You can use a blank form (typically provided by CMMS vendor, see Fig. 5.2) to collect equipment and PM data. A form like this will also ensure the right order of data fields. It should include the equipment number, manufacturer, model, location, serial number, PM procedure, frequency, and starting schedule.

Many equipment manufacturers are starting to develop libraries of information, for example, operating and maintenance manuals. Many vendors have this information available on line that you can link with your CMMS. This can help in initial data creation.

Labor. The next step is gathering information on maintenance employees. You need information on each maintenance technician, such

Input form: Equipment data

Equipment # ______

Description ______

Department ______

Location ______

Priority ______

Model # ______ Serial # ______

Purchase price ______ Purchase date ______

Installation date ______

Manufacturer ______

Vendor ______

Warranty date ______

Meter reading

Date		Reading	

Spare parts

Part number	Quantity

Preventive maintenance

Task description

Frequency ______ Days ______ Run time

Next due ______

Generate PM based on ○ Next due ○ Completion of PM

Figure 5.2 Equipment input form.

as name, address, phone, and social security. You probably have this information in the existing system or you might have to obtain it from HR or payroll department. You can use a blank form (typically provided by CMMS vendor) to collect this information.

Work order

Work request. Set up user IDs and passwords for all possible personnel who are likely to use the CMMS to generate work requests. Make sure it is easy to maintain and update this information.

A scheme for all codes to be used within work order system was designed during the planning phase. Now is the time to actually compile a list of all these codes.

Codes shown in Table 5.2 are some examples. There are a number of databases on this topic in existence. And there are many sources for information on these including trouble-shooting guides from manufacturers and an initial library of corrective actions either from manufacturers or from employee knowledge and skills.

TABLE 5.2 Code Tables

Work order status	
APPR	Approved
WTAPP	Waiting for approval
WM	Waiting for material

Work order type	
EM	Emergency
PM	Preventive maintenance
RM	Repair maintenance
PJ	Projects

Failure code	
BF	Bearing failure
LO	Low oil
WT	Wear & tear

Parts inventory. The next step is gathering information on the parts. You do not necessarily need to include all the parts in the beginning. You may start with critical parts and build rest of the database as you go along.

To collect parts data, most likely, a manual system exits with cards providing the part number and other limited bits of information. This is where you start. You need information such as part number, category, location, vendor, price, quantity-on-hand, and unit of measure.

You can use a blank form (typically provided by CMMS vendor, see Fig. 5.3) to collect parts data. A form like this will also ensure the right order of data fields.

The information-gathering activity requires considerable time and numerous calls to vendors to obtain all the detailed information essential for maintenance inventory. Remember, one reason many companies fail in implementing a CMMS is difficulties they experience at this point. A major shortcoming is failure to commit sufficient manpower for data gathering.

Input form: Parts inventory			
Item #			
Description			
Category			
Issue unit			
Manufacturer part #		Vendor part #	
Quantity on hand		EOQ	
Max quantity		Min quantity	
Account #		Price	
Location #1			
Location #2			
Location #3			
Vendor #1			
Vendor #2			
Vendor #3			
Where used			
Equipment #1			
Equipment #2			
Equipment #3			
Equipment #4			
Equipment #5			
Equipment #6			

Figure 5.3 Parts inventory input form.

The question that usually comes up is how many people are needed to gather this data? That depends on how fast you want to implement the system and how good your existing records are. For a system of 20,000 parts, you should have a full-time person collecting information. Depending on how good the manual records are, it may take two to six months.

Mass data requirements. This area in particular is becoming easier to facilitate in a rapid manner. Large data libraries are becoming more and more accessible. For example, many vendors are beginning to create online parts listings for their equipment. These can easily be transferred to the CMMS. Some CMMS can support live linking, further reducing the workload for this task. A database of such materials will cut short the implementation time significantly.

Data entry

The next phase is data entry after data gathering is completed. Basically you have three options:

- *In-house.* Use your own clerical personnel to do the data entry.
- *Outside temporary help.* Hire temporary people to work under your direction for entering the data into your CMMS. This is effective if you do not have committed manpower.
- *Outside contractor or consultants.* These people do a turnkey job of data gathering and entry. This decision would have been made during data gathering or planning phase.

Once the decision is made as to who is going to enter the data, the next step is to develop a prototype. Enter a few records into your CMMS and go through the whole cycle of generating work orders, completing work orders, generating work histories, and so forth. You will encounter problems and difficulties. Work with your CMMS vendor to resolve them. Keep repeating this process until everything is working to your satisfaction.

Enter all the data when you are fully satisfied the CMMS is working adequately and then "go live!"

Follow up/monitoring

Merely keeping the system running smoothly is not enough to justify its continuing existence. It also must produce something useful. To ensure that goal, make sure that the CMMS continues to serve the purpose for which it was purchased.

There are many different KPIs available for evaluating maintenance performance. Among them are maintenance dollars per pound of product, maintenance cost per company employee, and equipment maintenance hours per production hour. All of these KPIs contribute to the general measures of maintenance effectiveness, but no one KPI can monitor everything.

During the justification process, you had developed a list of KPIs for your application. Now is the time to monitor those to ensure that you are meeting or exceeding the goals.

Let us look at how to monitor and analyze KPIs. As an example, the total of maintenance hours completed must be compiled (probably normal outputs for most CMMS). The totals for each category (PM, repair, and the like), as a percentage of the total maintenance hours, may then be compared to monitor how well the system and the maintenance effort, in general, are performing.

If the unplanned work rises, two possibilities exist: there may be more breakdowns in general or there may have been a single large problem. Whatever the reason, the increase in unplanned work means less total maintenance work done by the workforce due to the lower productivity of unplanned time. A closer look at the CMMS summary reports and work orders will provide initial answers as to where the problem lies. If the problem stemmed from a single large failure, the maintenance effort may still be performing as needed. The failure should still be investigated to find its cause.

If breakdowns have indeed been increasing, the condition of the equipment or the effectiveness of maintenance work may be deteriorating. These possibilities must be explored.

Slacking off on PM results in an increase in the breakdowns that cause emergency or unplanned work. Therefore, monitoring of PM work is also a good index of maintenance system effectiveness and future equipment breakdown expectations. This is how CMMS acts as a valuable tool in enhancing your productivity.

Upgrading your CMMS. As system usage settles in and experience is developed, possibilities for improving the system become apparent. Discussions occur with suppliers and other users on how to make the improvements. The upgrade/replace process should follow the same general path as the original purchase process. First establish the need, develop support for the change, then find the system upgrade or replacement to accomplish the change and implement.

The time to consider an upgrade or replacement of your CMMS is when you become aware of the need for additional capabilities, which can be cost justified. This may occur, for example, through the persistent lack of data needed for making specific decisions or producing needed reports. Frequent occurrence of this situation is a sign that the system's needs were not fully understood during system selection, system usage has grown beyond original expectations, or some combination of the two. Whichever is true, system improvement or replacement is indicated.

Most systems have a great deal of growth potential built into them. Much of that potential can remain unused throughout the life of the system because the users are not aware, or have forgotten that it is there. A recent CMMS survey (see App. 6A) by the author shows that only 6 percent of companies are using their CMMS to its fullest capacity. In some cases, using those untapped resources can eliminate the need to upgrade or replace. The key is to know what your system's true capabilities are and use them as required. More details on this topic are covered in Chap. 6.

Conclusion

Follow a step-by-step process for successful implementation. Avoid mistakes that have led to majority of failures in the past such as lack of upper management commitment. Make sure you have appropriate resources for manpower and training. One of the crucial elements for implementation is planning. With proper planning, you can enjoy a CMMS success for a long time.

Appendix 5A Case Study: CMMS Implementation*

It may take a long time to implement a computerized maintenance system, but the effort yields an improved and efficient stock room with a quantifiable knowledge of all items in stock. According to company XXX, Canada's one of the largest dairy processors and a global cheese producer, management spent fruitful efforts to gather inventory data and number of each individual stock item. Now, XXX tracks part usage and saves cost on part maintenance.

Before implementation of a *computerized maintenance management system* (CMMS), the company lacked a valuable preventive maintenance operation, had no record keeping for *hazard analysis and critical control point* (HACCP) compliance, and no inventory control.

Between production planning, sales forecasting, inventory analysis, maintenance scheduling, and budgetary demands, there are many things to consider before going ahead with a CMMS.

---

*This case study is contributed by Julie Waterhouse.

Data gathering enables the right decisions to lay out a project plan, estimate time frames, and determine the resources needed to accomplish a successful CMMS. The bulk of management's implementation time was spent with actual data collection on everything from equipment information, inventory status, and instructions for preventive maintenance. First, it was important to build an equipment list, which did not exist at this plant and then enter the only major pieces of equipment. As time passed, the equipment list has been filled out with more components and it will continue to do so for quite some time.

In order to develop task instructions, management considered the basic instructions, information from equipment manuals, and the service technicians' experience.

It was a little more legwork as old photocopies were currently being used for maintenance tasks with no soft copy available. The basic instructions were typed into the task descriptions, along with some more substantial information from equipment manuals, and the service technicians' experience. Scheduling frequencies were also a combination of manufacturer's recommendations, their own experience, and regulatory procedures outlined by HACCP. Sanitary maintenance was much easier because they were able to import Excel spreadsheets of instructions into the database.

The next and most time-consuming data collection project was in the stock room. With no existing physical list of the items in stock, a lot of time was required to intake all the information about the part numbers kept in inventory and input it into the system. Once everything was entered, suppliers were contacted for pricing, minimum order quantity, and lead-time information. Upon receiving, this information was entered into the CMMS.

XXX chose MP2 by Datastream for their CMMS. Equipment records in MP2 have room for so much information, but the focus was mainly on model number, serial number, vendor, account code, and location in the plant. During implementation, XXX set up the nomenclature as 1496-department-type and number. The nomenclature defines the HACCP registration number, 1496, then the department where the equipment is installed, then the equipment type and a sequential number. For example, 1496-PKG-FILL01 is the first filler located in the packaging department. This nomenclature is highly configurable as it is easy to add more components to the nomenclature (i.e., 1496-PKG-FILL01-PMP01).

Upon implementation of the CMMS, new pick lists are created for preventive maintenance tasks and large scale repairs. One of the biggest bonuses from implementation is that the company no longer has to buy a lot of stock at the same time, because it is easy to track and trend part usage. CMMS can be utilized to save the time and expense of running out of stock.

Management has already grown fond of the CMMS in terms of its ability to promote and maintain compliance with HACCP, especially since the inspectors are familiar with CMMS and the record keeping has become automated, thorough and complete. XXX believes measured successes will enable this system to become a natural and reliable part of the asset management system throughout the plant. As the system ages, matures, and builds equipment history, CMMS will continue to gain the respect of the maintenance department, as well as the rest of the plant.

Chapter

6

How to Audit/Optimize Your CMMS

Introduction

Maintenance departments in school districts, universities, hospitals, government buildings, and commercial office buildings nationwide increasingly rely on a *computerized maintenance management system* (CMMS) to gather, sort, analyze, and report on essential information related to equipment and facilities performance. Managers use this information, among other things, to set department priorities and cost-justify equipment purchases. In many cases the CMMS is not producing the desired results. The question is, when is the time to upgrade your CMMS.

Let us understand the meaning of optimization and upgrading of a CMMS. *Optimization* means determining existing useful features in your CMMS currently not being used and start utilizing them to improve productivity. *Upgrading* means determining lack of useful features in your CMMS and then obtaining them either by upgrading your current CMMS or by acquiring a new CMMS package.

The most important step in the upgrade/optimize process is an audit of your CMMS. Observations based on audits reveal how audits can form the basis of a CMMS upgrade.

The dynamic nature of business operations and the continuous challenge to keep costs down makes periodic audits a necessity if the businesses are to succeed. Two major steps comprise the audit procedure. The first step is establishing a baseline, and the second is comparing subsequent audits to the baseline to measure improvements.

Essentially the audit shows strengths and weaknesses. The strengths are continued and the weaknesses are analyzed to establish actions for improvement. For long-range improvements the audits are required at least once a year to continue the improvements.

CMMS Audit

No.	Features	Importance Rating A (*1 through 10*)	Available But Not Using B (*Y or "Blank"*)	Not Available C (*Y or "Blank"*)	Score 1 A × B	Score 2 A × C
General						
1	The CMMS should be 100% web-based.				0	0
2	The CMMS must run on an industry standard relational database management system. For example, MS SQL, Oracle, and so forth.				0	0
3	Should support open source application.				0	0
4	Must have back up/archive functionality.				0	0
5	System can be integrated to purchasing, engineering, payroll/accounting, and the like.				0	0
Ease of Use						
1	The CMMS must utilize a context sensitive help menu that can be edited and appended by the end users to create an in-house knowledge base on their business practices.				0	0
2	The CMMS must utilize a query function to find data (full or partial match).				0	0
3	CMMS should be able to move directly from one screen to another via hyperlink, menu selections, buttons, or tabs without returning to a "main menu."				0	0
4	The CMMS must visually distinguish between fields that are required and fields that are displayed only.				0	0
5	CMMS must have ability to duplicate records and then modify as desired.				0	0
6	The system must allow users to customize system codes that appear throughout the system. Key information to be user-defined.				0	0

7	Reporting function: Must provide the users ability to easily modify existing reports and create new reports.	0	0
8	Ability to generate graphics reports.	0	0
Data Integrity and Security			
1	The CMMS must provide a means to set up user accounts, assign user passwords, and assign users desired access.	0	0
2	CMMS should provide security at the field level, i.e., hiding a particular field from others.	0	0
Asset Management			
1	Ability to attach multimedia objects (video, graphical files, or sound) directly to equipment and then viewed as needed.	0	0
2	Should be able to schedule preventive maintenance by calendar time or meter usage. Meter usage can be single or multiple (hours, temperature, miles, and so on).	0	0
3	The system must allow a sufficient amount of textual data entry for description of each equipment.	0	0
4	System should be able to suspend, inactivate, transfer, dispose, or retire an asset.	0	0
5	The system must fully incorporate the parent/child relationship to identify subassemblies of assets.	0	0

(Continued)

CMMS Audit

No.	Features	Importance Rating A (*1 through 10*)	Available But Not Using B (*Y or "Blank"*)	Not Available C (*Y or "Blank"*)	Score 1 A × B	Score 2 A × C
Preventive Maintenance						
1	The CMMS must allow preventive maintenance activities by a fixed method (for example, every month regardless of when the last preventive maintenance was completed).				0	0
2	The CMMS must allow preventive maintenance activities by a floating method (e.g., 30 days from the previous preventive maintenance close date).				0	0
3	The CMMS must have the ability to set up and schedule seasonal preventive maintenance activities.				0	0
4	CMMS should have the ability to link to a variety of resources such as scanned or CAD drawings, reference documents, graphics files, video files, and manufacturers' Web site.				0	0
5	CMMS should have the ability to recalculate preventive maintenance labor hour estimates based on task history of actual hours.				0	0
6	Suspend PM schedules when equipment placed out of service or retired.				0	0
7	The system should be able to handle route PMs such as lube route, and filter change route.				0	0
8	Allows mechanics to easily write up deficiencies found on PM inspection tours as planned work to be done and automatically generate a work order(WO).				0	0

Work Orders			
1	Work orders must have the ability to be distributed via e-mail, printers, text-enabled pagers, and other hand-held devices.	0	0
2	Should have the ability to set “warning” flags and print on WO. Link up health and safety procedures such as confined space, lock or tag out, protective clothing, or apparatus to be worn (all user definable).	0	0
3	Ability to apply corrections to closed and completed WOs.	0	0
4	Should allow for both automatic and manual WO numbering.	0	0
5	Ability to incorporate bar code and PDA technology.	0	0
6	Ability to manage multiple workers on one WO.	0	0
7	Ability to allocate parts to a WO.	0	0
8	Compare budgets or estimates against actual data and provide exception reports.	0	0
9	Task library of common work instructions.	0	0
10	PM/safety procedures should be printed on the WOs.	0	0
11	Provides status of all outstanding WOs.	0	0
12	Ability to generate equipment/asset history from birth (installation, construction, or connection) with all repairs.	0	0
13	System reports on contractor versus in-house work.	0	0
14	Provides reports charging back maintenance cost to department or cost center.	0	0
15	Has reports with mean time between failures that show how often the unit has been worked on, how many days (or machine hours) lapsed between failures, and the duration of each repair.	0	0

(Continued)

CMMS Audit

No.	Features	Importance Rating A (*1 through 10*)	Available But Not Using B (*Y or "Blank"*)	Not Available C (*Y or "Blank"*)	Score 1 A × B	Score 2 A × C
Inventory						
1	Ability to handle multiple stock rooms.				0	0
2	Track warranty.				0	0
3	Track stocked/nonstocked, Taxable/nontaxable, and rebuilt items.				0	0
4	Status reports and notification of WOs awaiting parts.				0	0
5	Multiple vendors per part, multiple order options per vendor.				0	0
6	Multiple manufacturers per part.				0	0
7	"Where used" option; that is, listing of all equipment where a certain part number is used.				0	0
8	Ability to handle bar-code enabled counting.				0	0
9	Backorder and partial shipment processing.				0	0
10	Repairable spare tracking.				0	0
11	Allow transfers between stock-rooms.				0	0
12	Provision for identifying inactive material.				0	0
13	The ability to track items by serial number.				0	0
14	Flag for withdrawal/usage of a part that is higher than normal.				0	0
15	Track changes in prices.				0	0
16	Provision for physical count.				0	0
17	Allow purchasing and issuing in different units of measure.				0	0
18	Allow blanket orders.				0	0
19	Allow automatic suggestions for reorder points based on usage.				0	0
20	Receive and withdraw by both bar code and manually.				0	0
21	The system must have the capability of creating "kits" of commonly used parts to reduce inventory shortages.				0	0

Planning and Scheduling			
1	Account for WO and equipment priorities.	0	0
2	Ability to import/export data to production scheduling system.	0	0
3	Provision to define planning periods.	0	0
4	Work order schedule to be generated automatically with manual override capability.	0	0
5	Accommodate shutdown schedules.	0	0
6	Craft available and required man-hours comparison.	0	0
7	Utilize backlog management for resource leveling.	0	0
8	Match task to the talent or skill of a worker.	0	0
9	Use various forecasting reports to assist in planning and analyzing availability for PM work, special project to be completed.	0	0
Purchasing			
1	Purchasing.	0	0
2	Ability to interface with existing purchasing system.	0	0
3	Autogenerate requisitions based on ROP.	0	0
4	Provision for autoapproval requisitions based on predefined criteria with manual override capability.	0	0
5	Autogenerate POs for all approved requisitions.	0	0
6	Provision for blanket POs.	0	0
7	POs to be generated to the primary vendors with capability to change it to alternate vendors.	0	0
8	Full or partial receipts.	0	0
9	Parts receipt manually or using bar-code technology.	0	0
	Total =	0	0

How to Use This Audit

- The audit provides a list of features for various CMMS modules. This is not a list of features that goes into your request for proposal (RFP). This is a list of some selected key features. You can add or delete based on your application.
- The next column is "importance rating (A)." Enter a number from 1 to 10; 10 being most important to your application and 1 being least important. Enter 0 if a feature is not applicable to you. For example, one of the features in inventory module is "ability to handle multiple stock rooms." If you have only one stock room with no plans in the future to change that, you would enter a 0 for importance rating (A).
- The next two columns are "available but not using (B)" and "not available (C)." Carefully review each feature. Enter a Y under B if you have a feature in your CMMS but are not using it for whatever reason and leave C as blank. Enter a Y under C if a feature is not available in your CMMS and leave B as blank. For scoring purpose, Y = 1, and a blank = 0.
- The next two columns "score 1" and "score 2" will be automatically computed.

Audit Results

Score 1 indicates the need to *optimize* your CMMS. Score 2 indicates the need to *upgrade* your CMMS. Please review Fig. 6.1.

- If score 1 and score 2, both, are Low; it is an ideal situation. This indicates that your CMMS is working for you. If score1 is high and score 2 is low; it indicates you need to optimize your CMMS. In other words, your CMMS has the features you need, except they are not being used currently. You need to contact your CMMS vendor for help. May be you need additional training or you need to take another look at implementation. May be an outside consultant can help in the process of proper implementation.

  If score 1 and score 2, both, are high; it indicates need to optimize and upgrade. First, follow the guidelines mentioned earlier for optimizing your CMMS. Before you decide to buy a new package, work with your existing vendor. Perhaps you are using an older version and the newer version might have many of the features you are looking for or the vendor might be able to do custom modifications to accommodate your requirements. Try these first and then perform another audit. Make a decision based on this audit.
- If score 1 is low and score 2 is high; it indicates a need to upgrade your CMMS. In other words, your CMMS does not have the desired

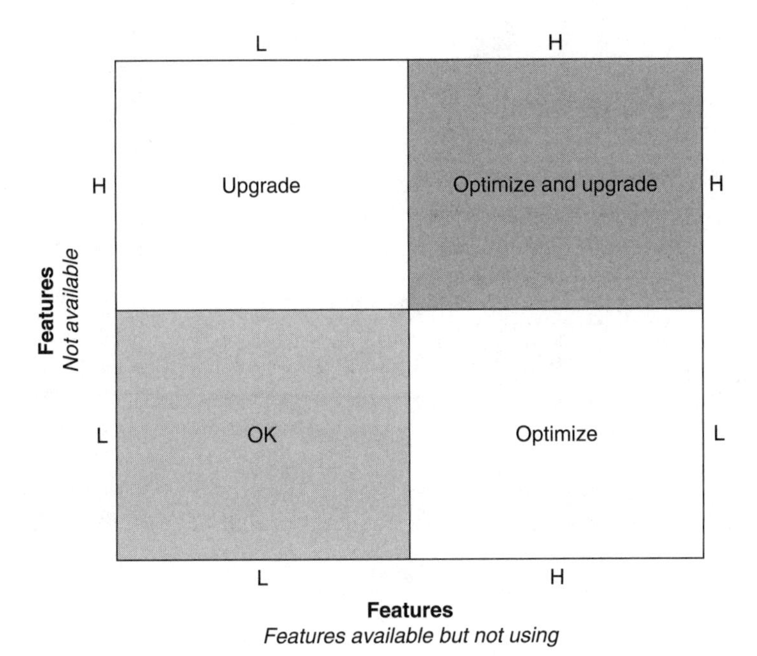

Figure 6.1 Audit results.

features. If your vendor cannot customize existing package in a reasonable time and for a reasonable cost; you are left with the only option to upgrade to a new CMMS.

A CMMS survey was conducted by the author in "Fall 2005." Approximately 300 people responded. A summary of responses and results is shown in App. 6A.

Conclusion

Whether upgrading to a new CMMS or optimizing an existing one, go through the audit of your current CMMS to determine the appropriate plan of action for your situation. Studying why so many CMMS installations have failed and following the proper selection process for your CMMS will help you acquire a CMMS, which will work for you for years to come. Otherwise, the latest and greatest technology will yield no better results than the old system did.

Appendix 6A CMMS Survey

Following are some of the comments by respondents:

1. What is the primary business at your work location?	Response Percent	Response Total
Manufacturing	32.1%	96
Process Industry	2%	6
Construction	1.7%	5
Education / Training	10.4%	31
Hospitality / Travel	2%	6
Healthcare / Pharmaceutical	10.7%	32
Consulting / Services	6%	18
Utility / Communications	4%	12
Government	9.4%	28
Transportation	2.3%	7
Non-profit	1.3%	4
Wholesale / Retail	2.7%	8
Other (please specify)	15.4%	46
	Total Respondents	299

2. What is the total number of equipment pieces/assets that are maintained under the CMMS system?

Total Respondents | 244

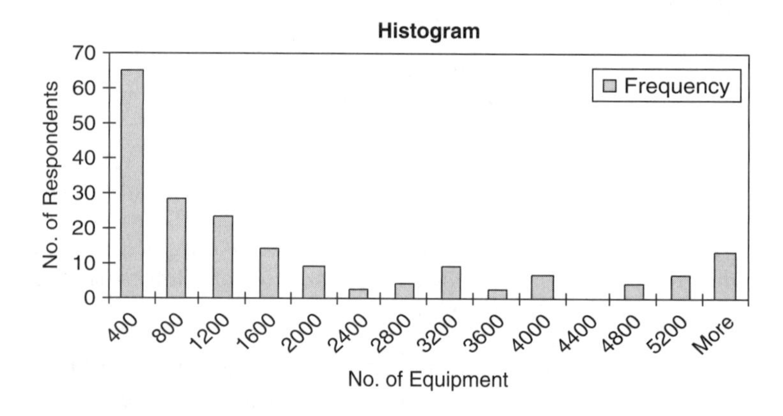

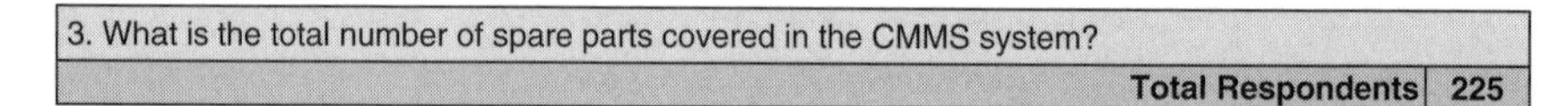
3. What is the total number of spare parts covered in the CMMS system?

Total Respondents | 225

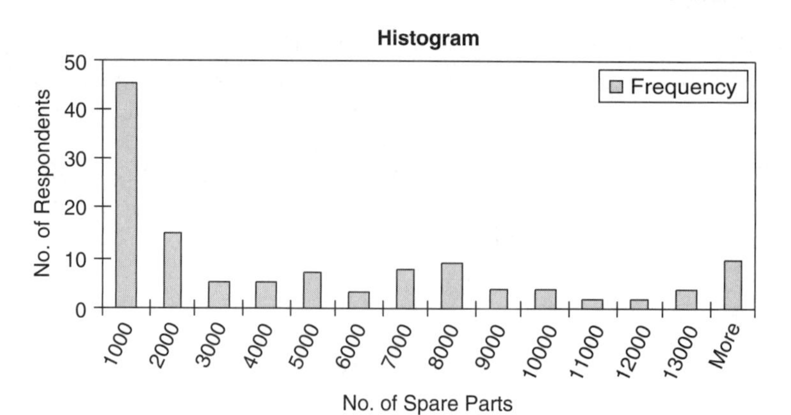

4. On average, how many work orders are issued per week?	
Total Respondents	**245**

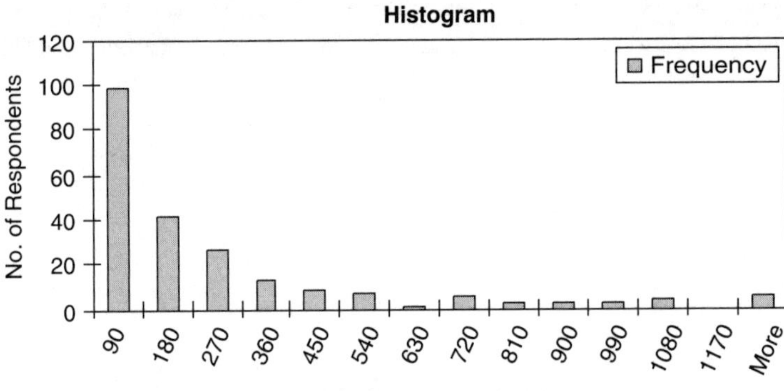

5. On average, how many purchase orders are issued per week?	
Total Respondents	**233**

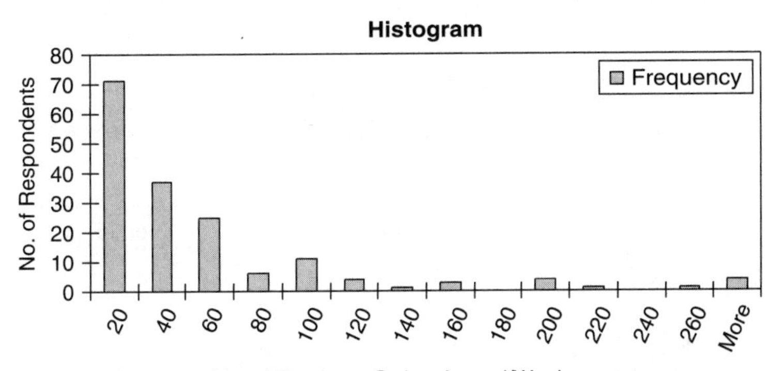

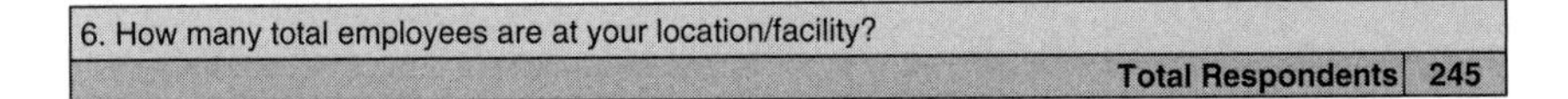

6. How many total employees are at your location/facility?	
Total Respondents	**245**

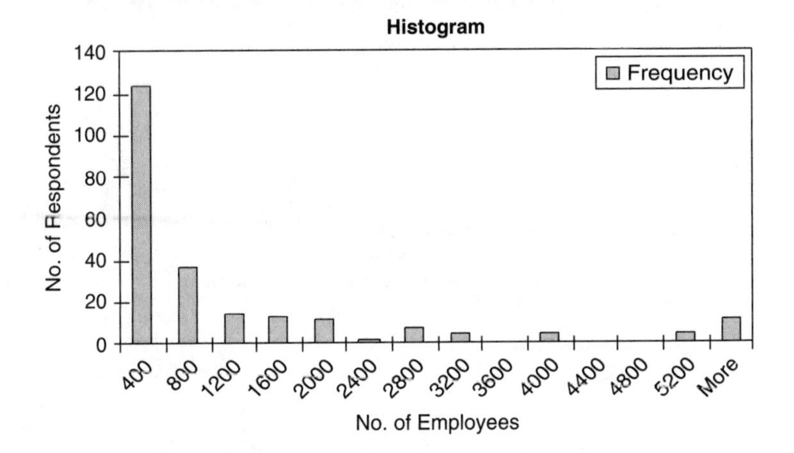

7. How many manufacturing production employees are at your location/facility?	
Total Respondents	**233**

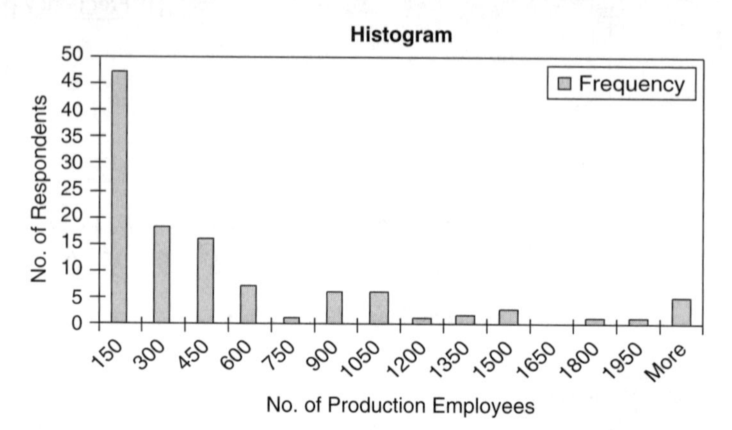

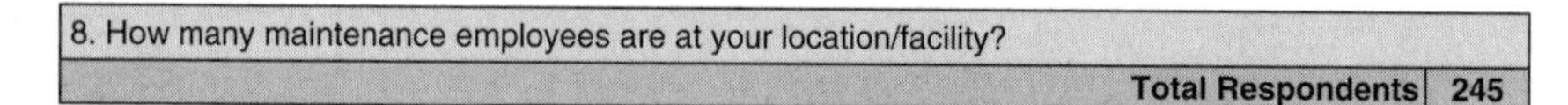

8. How many maintenance employees are at your location/facility?	
Total Respondents	**245**

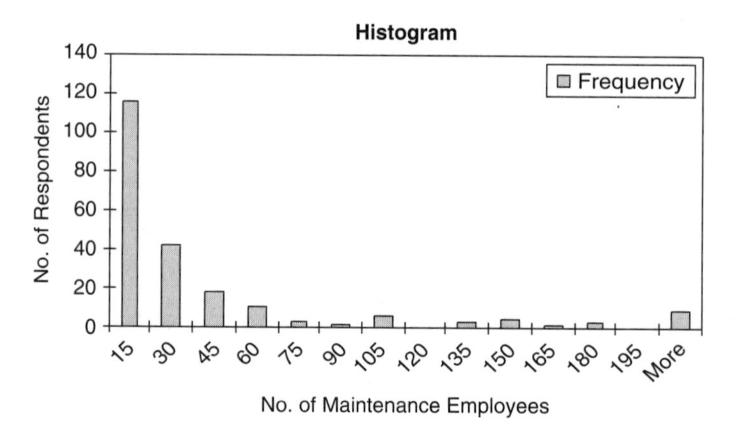

9. Which CMMS modules do you currently use?

	Response Percent	Response Total
Equipment	78%	184
Preventive maintenance	**86.4%**	**204**
Work order	83.5%	197
Inventory	43.2%	102
Purchase order	29.7%	70
Other (please specify)	22%	52
	Total Respondents	**236**

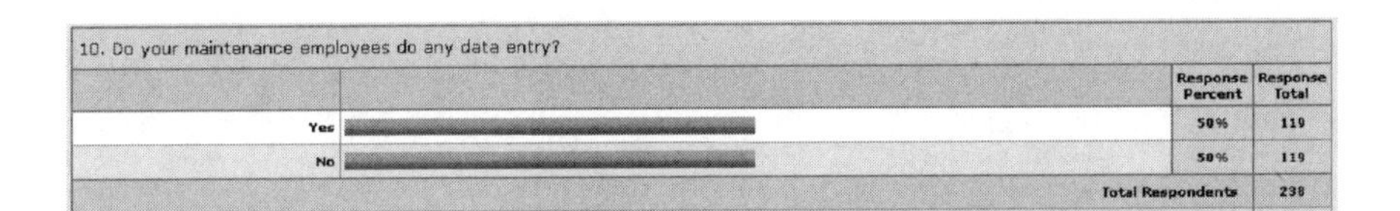

10. Do your maintenance employees do any data entry?			
		Response Percent	Response Total
Yes		50%	119
No		50%	119
		Total Respondents	238

11. On average, how many hours per week do your maintenance employees spend on data entry?	
Total Respondents	117

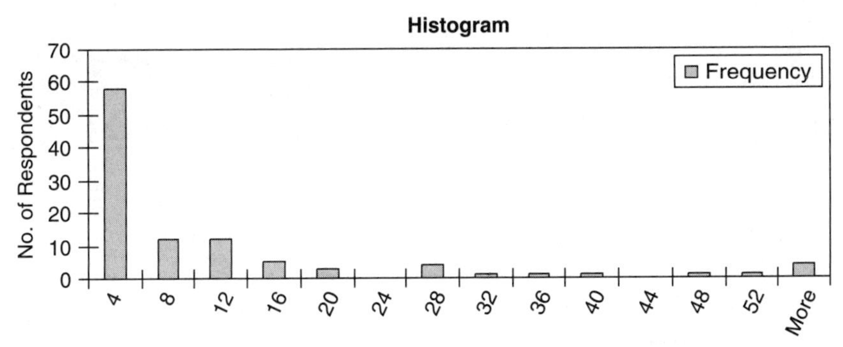

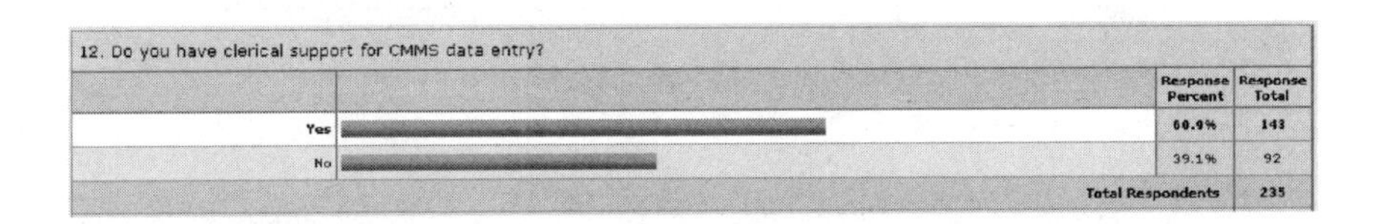

12. Do you have clerical support for CMMS data entry?			
		Response Percent	Response Total
Yes		60.9%	143
No		39.1%	92
		Total Respondents	235

13. On average, how many hours per week does the clerical support staff spend on CMMS data entry?	
Total Respondents	138

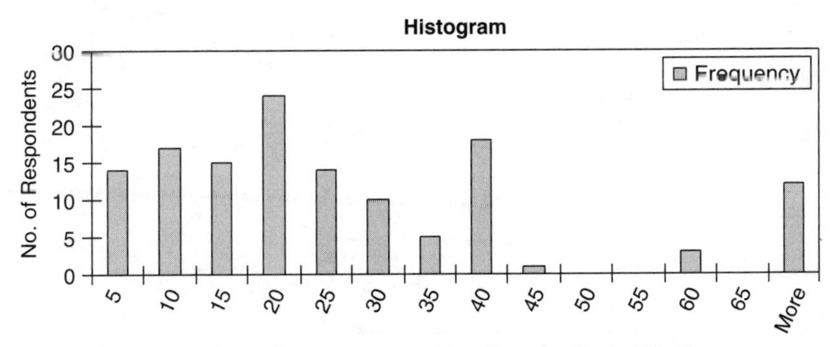

14. How long did it take your organization to implement CMMS?	Response Percent	Response Total
1-3 months	8.9%	20
4-6 months	17%	38
7-9 months	8.5%	19
10-12 months	15.2%	34
1-2 years	16.1%	36
2-3 years	12.9%	29
Other (please specify)	**21.4%**	**48**
	Total Respondents	**224**

15. Did you use maintenance employees to collect data for your CMMS implementation?	Response Percent	Response Total
Yes	**78.1%**	**178**
No	21.9%	50
	Total Respondents	**228**

16. Approximately how many total hours did the maintenance employees spend gathering data for the CMMS implementation?	Response Percent	Response Total
100-200 hours	**28.4%**	**48**
201-300 hours	11.2%	19
301-400 hours	10.1%	17
401-500 hours	9.5%	16
501-600 hours	5.3%	9
601-700 hours	3.6%	6
701-800 hours	13.6%	23
Other (please specify)	18.3%	31
	Total Respondents	**169**

17. Did you have an outside contractor gather data for the CMMS implementaion?	Response Percent	Response Total
Yes	17.8%	38
No	**82.2%**	**176**
	Total Respondents	**214**

18. Did you use an outside contractor to enter data?	Response Percent	Response Total
Yes	17.2%	37
No	**82.8%**	**178**
	Total Respondents	**215**

19. Does your organization hire interns or co-op students to work in the maintenance department?	Response Percent	Response Total
Yes	30.6%	66
No	**69.4%**	**150**
	Total Respondents	**216**

20. Every organization spends a certain amount of money on continuing education (such as training seminars and conference attendance). What percent of your organization's total training budget is allocated for maintenance training?

		Response Percent	Response Total
1%		38.3%	82
2% - 4%		30.4%	65
5% - 10%		10.3%	22
11% - 15%		7.5%	16
Other (please specify)		13.6%	29
		Total Respondents	214

21. How often does a machine (or other capital asset) go down and the part or parts needed to fix it are unavailable?

		Response Percent	Response Total
More than once a day		2.3%	5
Daily		5.6%	12
Twice a week		7%	15
Weekly		18.8%	40
Monthly		31.5%	67
Other (please specify)		**34.7%**	**74**
		Total Respondents	213

22. Is your upper management committed to the CMMS program?

		Response Percent	Response Total
Yes		**77.5%**	**165**
No		22.5%	48
		Total Respondents	213
		(skipped this question)	86

23. How long did it take you to justify a CMMS program for your organization?

		Response Percent	Response Total
Less than 2 months		**28%**	**59**
2-4 months		8.1%	17
4-6 months		12.3%	26
7-9 months		5.7%	12
10-12 months		19%	40
Other (please specify)		27%	57
		Total Respondents	211

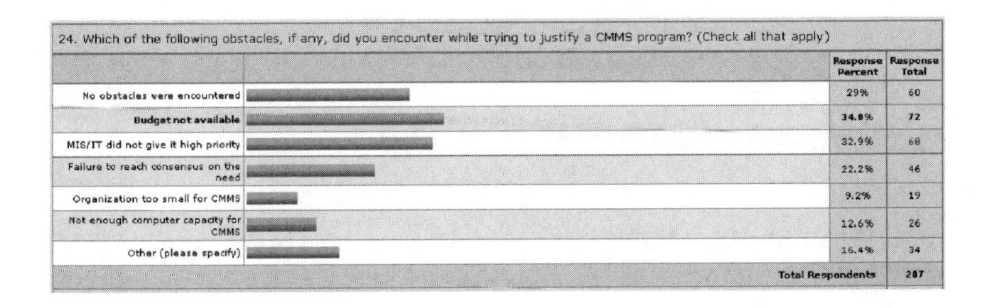

24. Which of the following obstacles, if any, did you encounter while trying to justify a CMMS program? (Check all that apply)

		Response Percent	Response Total
No obstacles were encountered		29%	60
Budget not available		**34.8%**	**72**
MIS/IT did not give it high priority		32.9%	68
Failure to reach consensus on the need		22.2%	46
Organization too small for CMMS		9.2%	19
Not enough computer capacity for CMMS		12.6%	26
Other (please specify)		16.4%	34
		Total Respondents	207

25. What percentage of your total annual maintenace budget are you saving as a result of better scheduling with the CMMS program?	Response Percent	Response Total
5% or less	28.6%	59
6%-10%	19.4%	40
11%-15%	15.5%	32
16%-20%	7.3%	15
Other (please specify)	29.1%	60
	Total Respondents	206

26. What percentage of your total annual maintenance budget are you saving as a result of increased parts availabiliy with the CMMS program?	Response Percent	Response Total
5% or less	40.7%	81
6%-10%	12.1%	24
11%-15%	10.1%	20
16%-20%	4.5%	9
Other (please specify)	32.7%	65
	Total Respondents	199

27. What percentage of your total annual maintenance budget are you saving as a result of increased machine availability with the CMMS program?	Response Percent	Response Total
5% or less	32%	64
6%-10%	19.5%	39
11%-15%	8.5%	17
16%-20%	9%	18
Other (please specify)	31%	62
	Total Respondents	200

28. What percentage of you total annual maintenance budget are you saving as a result of better inventory control with your CMMS program?	Response Percent	Response Total
5% or less	32.8%	65
6%-10%	15.7%	31
11%-15%	5.1%	10
16%-20%	4.5%	9
21%-25%	4%	8
26%-30%	2%	4
Other (please specify)	35.9%	71
	Total Respondents	198

29. Do you think you are using your CMMS to its maximum capability?	Response Percent	Response Total
Yes	5.3%	11
No	94.7%	198
	Total Respondents	209

30. If not using your CMMS to maximum capability, why not?		
	Total Respondents	177

- It is an old system that needs an upgrade.
- Software limitations.
- Lack of expertise to learn and maintain the system, and lack of clerical help for inputting data. The data has many capabilities but we do not have enough resources to utilize them.
- The program being used is too complicated.
- We are understaffed. In this day of "doing more with less" we have to focus our efforts where the biggest bang for our buck is, that is, PM and work order control.
- The CMMS has modules [parts inventory, *purchase orders* (POs)] that are not currently being used. The reason is the time and data collection needed to load the module.
- The PO system in place is a corporate mandated system controlled by the central business office, so we cannot use the POs generated by our system.
- More modules than we have the capacity to use. Our limitation is data entry and clerical staff.
- It would take precious time from the maintenance effort. Not just hours sitting at a machine, but training time learning more about the program so that new information can be used more effectively.
- Not enough resources are allocated to utilize what is available to us.
- Lack of training for the staff; a long time would be needed to train them on all the capabilities of CMMS.
- Not enough resources, maintenance staff not computer literate.
- Training, dedication, staffing, background of maintenance staff varies significantly from facility to facility. Have not been successful in establishing absolute company standards for usage.
- Poor interface with accounting and SAP software. Double entry of all POs.
- Not using data to make decisions. CMMS is more of an automated record keeper than an analysis tool.
- System is dated and does not have the necessary capability to have integrated links with inventory, financial, timekeeping, and human resource.
- Lack of technical ability of staff to understand the system, lack of managerial support, time, system support, and lack of upgrade budget dollars.
- Resistance to change.

- Still encounter too many stock outs. Have too much obsolete inventory. Data is not analyzed often enough to benefit from history acquired.
- The products that we have used are far too cumbersome and complex for our needs. The basic input and reporting functions are too complicated for an average staff member to access and utilize.
- Undersized maintenance staff for facility size, so we must apply our limited resources to putting out fires.
- I only have five people including myself and six separate storage areas for parts and materials. To implement inventory section, I have to get a grip on my inventory and I have no time to do this task. Purchasing is done by me through our purchasing department (75 percent of the time) and they do not want to be associated with our CMMS. In fact none of the other sections (including finance) want to be associated with CMMS. The thought that maintenance is capable of having a brain and excellent management abilities is a foreign concept.
- My organization does not understand the maintenance process. The traditional focus has been on asset management focused on revenue generation. Deferred maintenance has been standard operating procedure in the organization.
- Lack of interest within the entire organization from labor to management. The management of the organization had buried their heads in the sand for many years when dealing with the needs of anything related to the facility. The attitude was, if it does not make direct money, do not do it. Only expenses directly related to product going out the door were considered.
- Lack of training and reluctance to change from reactive mentality to proactive maintenance philosophies: "if it isn't broke, why am I spending dollars on it?"
- Little support from IT group.

31. Do you use any handheld devices (such as bar code system, palm, etc.) with your CMMS?

	Response Percent	Response Total
Yes	22.9%	49
No	77.1%	165
Total Respondents		214

32. Please explain the applications of the handheld devices with your CMMS.

Total Respondents	46

Following are some other comments by respondents:

- Monthly fire extinguisher, exit sign, and fire hose location checks.
- Inventory control/bar-code reading when pulling materials/parts from stock—resulting in automatic reordering and precise maintenance of Kan Ban "min/max" levels. Vibration analysis—hand-held data collector plugs into vibration monitor to capture data, and then plugs into PC for downloading.
- Inspections, auditing.
- Our maintenance techs have one to download their work orders and track their labor and materials.
- Stockroom inventory control.
- Bar-code readers for part issues and charge-outs, part receiving.
- Tricoder hand-held scanner, for parts receiving, parts inventory, parts usage applied to work orders, truck wash module, fuel module, update hour reading (mileage).
- Remote work requests can be entered into palm.
- Wireless hand-held devices capable of identifying work requirements during inspection process.
- Real time PDAs for work orders.
- Inventory cycle counting.
- Handhelds contain WOs—so the system is paperless.

33. Did you consider developing a CMMS package in-house?

	Response Percent	Response Total
Yes	34.8%	70
No	65.2%	131
Total Respondents		201

34. What is your ratio of planned work orders to emergency work orders?

	Response Percent	Response Total
100 planned to 1 emergency	10.6%	22
100 planned to 5 emergency	**22.1%**	**46**
100 planned to 10 emergency	13.9%	29
100 planned to 15 emergency	5.3%	11
100 planned to 20 emergency	12%	25
100 planned to 30 emergency	4.3%	9
100 planned to 50 emergency	9.1%	19
100 planned to 75 emergency	7.2%	15
Other (please specify)	15.4%	32
Total Respondents		208

35. Of your total work orders, what percent are the PM work orders?	Response Percent	Response Total
1%	2.8%	6
5%	6.6%	14
10%	5.7%	12
15%	7.1%	15
20%	19.3%	41
50%	28.8%	61
Other (please specify)	29.7%	63
	Total Respondents	212

36. Do you use a work order priority system?	Response Percent	Response Total
Yes	78.4%	167
No	21.6%	46
	Total Respondents	213

37. Do you use an equipment priority system to indicate the relative importance of equipment?	Response Percent	Response Total
Yes	55.7%	117
No	44.3%	93
	Total Respondents	210

38. Do you use an imaging option (attaching CAD, digital pictures, scanned images to records, etc.) with your CMMS?	Response Percent	Response Total
Yes	26.2%	55
No	73.8%	155
	Total Respondents	210

39. Please explain the benefits of using an imaging option with your CMMS.	
Total Respondents	45

Following are some of the comments by respondents:

- Images, documents, lengthy instructions, and so forth, can be attached to each work order, bettering the information our technician has in the field. This information is then stored within our database and can be used for future tasking.

- Verifies what is to be worked on. Provides extra data such as prints or schematics.
- We normally attach a drawing or photo with a work order when needed.
- Scans in detailed PM procedures showing part locations pictures from manuals. Also includes safety procedures and line diagrams. Creates folders and all print outs as a packet.
- Contains historical data, parts ordering accurate information—saves time when looking up equipment information for rebuild or repairs—this is the latest thing we are getting into as more and more vendors have these “soft copies.”
- Electrical and hydraulic schematics, parts list, and equipment breakdown photos were also scanned into the system. This allowed a technician to view the information at the equipment location if a computer was available. Saved hundreds of trips and “idle” nonproductive time.
- To show vendors/contractors what the project is and the problem it is having.
- We keep reports such as oil analysis, and vibration analysis attached to the equipment history.
- Actual scanned documents and procedures from equipment manuals so that the mechanic has a complete understanding of proper procedures.
- PM and safety procedures are attached to equipment and work orders to help define procedures.
- Identification and annual condition of equipment.
- It is sometimes hard to recall just where the machine is located and which parts need close attention. Pictures are great for both of these.
- Data tends to get lost when put in folders or files. Having it attached to the equipment record makes it available all the time.
- Pictures say a thousand words: it is easy to justify the replacement of equipment for capital funding.
- Gives the technician the ability to access manuals and floor plans, as-builts and red lines from any computer terminal and eliminates the lost print factor.
- One of the facilities includes portions of the riser diagrams and critical components of the O&M manual of the particular equipment for quick reference. Such diagrams are then available on the palm.

- Standard operating procedures (SOPs) and safety information are available with each work order.
- Having prints of utilities or equipment ducts, piping, schematics, and the like eliminates trying to locate these during an emergency or rush job.

We would like to thank the following for participating in this survey:

- *Association for Facilities Engineering* (AFE)
- *Mapcon Users Group* (MUG)
- Reliable Plant Magazine

Chapter 7

Computerized Maintenance Management System and Total Productive Maintenance

Total productive maintenance (TPM) is defined as a strategy that introduces elements of a good maintenance program to increase overall equipment effectiveness and improve manufacturing processes.

The five key elements or "pillars" of TPM are:

1. Improving *overall equipment effectiveness* (OEE) by targeting the major causes of poor performance. Causes of poor performance include the equipment, parts, supplier, and individual performance.
2. Involving operators in the routine maintenance of their equipment.
3. Improving maintenance efficiency and effectiveness.
4. Improving skills and knowledge training.
5. Designing for operability and maintainability.

It is important to realize how *computerized maintenance management system (*CMMS) supports the key elements of TPM.

1. *CMMS has information that will calculate the OEE in order to determine improvement needs.* The OEE formula accounts for availability, performance, and quality.

$$\text{OEE} = \text{availability} \times \text{performance} \times \text{quality}$$

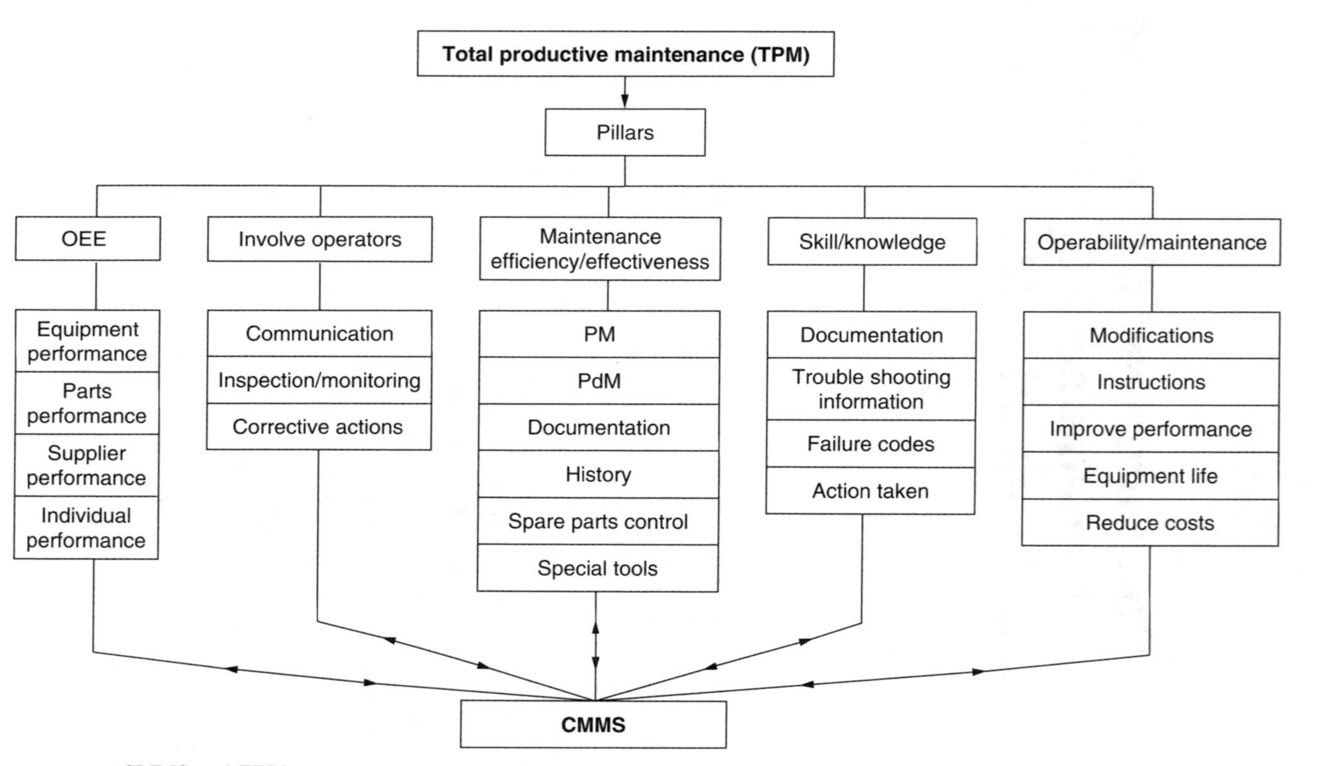

Figure 7.1 CMMS and TPM.

Equipment/asset downtime is monitored in many different ways. Manually, via PDAs or by machine monitoring devices. This information is fed into CMMS. Similarly, equipment performance and quality rate of each equipment is fed into CMMS.

2. *Involving operators in the routine maintenance of their equipment.* Whether machine operators do the routine maintenance or maintenance technicians do it, you still need to track who did the work, when, and material. CMMS will keep track of this maintenance history, and accordingly, generate the necessary work orders with all details of procedure, parts needed, and the like.
3. *Improving maintenance efficiency and effectiveness.* This can be done by proper preventive and predictive maintenance, equipment documentation (safety, schematic diagrams, and so on), repair and maintenance history, spare parts control, and so forth. All this is stored and tracked by CMMS.
4. *Improving skills and knowledge.* CMMS improves skills and knowledge with documentation of all procedures, trouble-shooting history, and equipment history (such as failure codes and corrective action).
5. *Designing for operability and maintainability.* CMMS enables modifications that extend part life, easier maintenance and inspection, improved performance, and extended equipment life while reducing costs.

It is evident that a tool such as CMMS is needed to support TPM philosophy (see Fig. 7.1). Author has shown this in relations to TPM. Similar analysis can be performed for other philosophies such as reliability centered maintenance (RCM).

Chapter

8

How to Turn Maintenance into a Profit Center

The Maintenance Image

The word "maintenance" conjures negative connotations in most people's minds. Frequently, when asked the question "what does the word 'maintenance' mean to you?" people respond with negative examples such as:

- A machine has broken down.
- There is some kind of problem with the equipment or the facility.
- Something needs to be repaired or fixed.

"These are all negative images, none of which reflect the true meaning of the word 'maintenance,' which is defined as '*the work of keeping something in proper condition; upkeep*'." The perception of the word "maintenance" has changed drastically; it has been twisted and pushed into an ugly light.

Ask yourself these questions:

- When expenses are cut, what goes first?

  Typically maintenance resources are cut first because upper management views maintenance as an expense. They just look at profit and loss statements and cut expense items to increase profits without realizing the consequences. We have all heard the term downsizing. As an example, if a company decides to downsize its operation by 20 percent, they almost always cut maintenance resources by 20 percent. Maintenance resources are based on company's assets. When you downsize, did you downsize the assets? In other words, did you get rid of 20 percent of your production equipment, or 20 percent of your buildings? No—that means maintenance now has to maintain more with less resources. This can have serious impact on a company's long-range survival.

- How many senior managers come from maintenance?
  I have asked this question to audiences for over 10 years. I have yet to see one hand go up. It shows the lack of emphasis and importance placed in the maintenance department.
- How many companies support maintenance R&D?
  Again, I have yet to see one. It goes back to the philosophy of how maintenance is viewed. Other departments or areas receive R&D funding as they are viewed as profit contributors. Millions of dollars are spent on product improvement R&D; sadly, it is forgotten that maintenance is an equal contributor to product quality improvement. You can spend millions on product design and quality improvement; but it will not be fruitful if you do not take care of the production equipment.
- How many interns are hired for the maintenance department?
  It would be ideal for a maintenance department to hire some interns to help implement a *computerized maintenance management system* (CMMS) project. A CMMS project involves a fair amount of data gathering and data entry. In fact, lack of resources to accomplish this part of the CMMS project is one of the primary reasons for CMMS implementation failures.
- How many maintenance courses do universities offer?
  Actually, there are two questions here. How many courses and how many universities? The answer to both is very few. This shows maintenance is not even viewed as a potential career for future graduates.

Cost Center versus Profit Center Approach

In order to truly see the positive effects that proper maintenance has upon an industry, we must view maintenance as a positive rather than a negative. Maintenance is a *profit center* not a *cost center*. Let us review the differences between a cost center and a profit center approach.

A cost center is concerned solely with controlling adherence to a budget. A cost center approach expects you to comply with an operating cost budget that resides below the gross profit line on an income statement. If you are restricted to managing budgets, why spend money on improvements or opportunities that may have a large impact on profitability with an increase in expenses?

Instead of encouraging efficiency and optimization, the cost center approach works against optimization. The amount under budget is added to the planned reduction and the result becomes next year's objective or in some cases money is allocated to other departments such as production. That is why there is so much last minute spending in a cost center. The reward works against optimizing, doing exceptionally well, and ending up significantly under budget.

A cost center approach treats maintenance as a charity. Because profit is the measure of success in a profit center, investment and operating costs can be allocated to improve efficiency. This allocation can result in unexpected opportunities to sell more and/or produce higher quality products. In a profit center, managers have the authority to reallocate existing resources as well as expend additional resources with corresponding accountability for results. A profit center mentality demands innovation and creativity.

How to Turn Maintenance into a Profit Center

Maintenance applications are everywhere and apply to every industry, whether it is manufacturing, metal fabrication, machine shop, casting, foundry, aerospace, food processing, chemical processing, electronics, healthcare, buildings and grounds, or utilities. It is the backbone of any operation and yet it is continually overlooked and neglected.

Let us examine how you can turn maintenance into a profit center.

Example No. 1 The term "product quality with zero error" is a common phrase among many industries. Essentially what this means is that all concentration and efforts are placed on the quality of the item without regard to the process or machinery involved in the making of the product. Seems a little bit backward, does not it? Professional athletic teams train throughout the year in order to maintain the athleticism and shape of their athletes, ensuring the best performance possible at time of competition or in our words "time of production." In terms of machinery and facilities the same should be true; keep machinery and facilities in maximum working order and the item produced will be the best possible, making the dream of "product quality with zero error" a much easier one to attain.

Producing higher quality products yields reduced return rate, thereby increasing your profits. That is one way you can turn maintenance into a profit center.

Example No. 2 *Overall equipment effectiveness* (OEE) consists of three factors: *availability, utilization, and quality rate.*

Availability. The percentage of time that the machine is available for production. *World-class standards demand 90 percent availability.*

Utilization. Also referred to as "design specs" or "performance efficiency." Essentially, it is the rating of the machine. The manufacturer provides the customer with a design specification rating for the machine. When the customer purchases the machine, the design specifications indicate machine output, for example, 200 pieces per hour. *The world-class standard for utilization is 95 percent, minimally.*

Quality rate. How good the final product is? Out of every 100 items produced, how many of them meet company standards of approval for distribution or sale? *World-class standards demand that companies have 99 percent quality rate, or 1 imperfect item for every 100.*

$$\text{OEE} = \text{availability} \times \text{utilization} \times \text{quality rate}$$

Based on the percentages noted earlier the equation appears as follows:

$$\text{OEE} = 0.90 \times 0.95 \times 0.99 = 0.85 \text{ (or 85\%)}$$

What that means is that per world-class standards, OEE for all machinery or assets should be at a minimum of 85 percent. North American companies average an OEE of only 40 percent. This is less than half of what world-class standards deem acceptable.

The most common remedy among North American companies for this problem is the purchase of more equipment with the incorrect assumption that this will lead to higher production levels. This is the wrong way to look at the problem. Bringing on more machines in order to compensate for the inefficiency and inconsistency of the existing, under-maintained machines is like trying to plug a hole in your boat with bubblegum. . . .great, if you only need it to hold for a few minutes, but terrible if it is expected to hold forever. More machinery means more potential for breakdown, especially when not maintained, and breakdowns affect everyone! Proper use of the OEE will provide companies with the greatest return on their assets. It will show how improvements in changeovers, quality, machine reliability improvements, and working through breaks will affect the bottom line.

Striving toward world-class productivity levels in any facility can seem like a daunting task, but this formula provides an easy benchmarking tool. The derived OEE percentage (85 percent) is easy to understand, and simply displaying it where all facility personnel can view it makes for a great motivational technique. It acts as a constant reminder of what the employees are striving for and what number they need to beat in order to stand ahead of the crowd.

Higher OEE means higher machine capacity, which in turn means higher output leading to increased sales capacity. This is a good example of how maintenance can be turned into a profit center.

On a larger scale, do not limit yourself to just equipment OEE, try to calculate production line OEE, and going a step further, try to even calculate the OEE of the entire facility.

Example No. 3 We all know that *preventive maintenance* (PM) can minimize breakdowns. Hundreds of thousands of dollars can be lost per hour during downtime. Hypothetically, let us say that for every hour of downtime, $1000 is lost. It is not uncommon for equipment to be down for a few weeks each year, let us say 100 hs per year. In this example, the company will lose $100,000 in the 100 hs of downtime. Multiply that by the number of machines per plant or facility, and the number could be staggering. If appropriate PM was in place, downtime would be minimized and tremendous amounts of money could be saved, which is exactly how *maintenance is turned into a profit center.*

How can a CMMS help?

With constant downsizing, maintenance has to produce more with less. How does a maintenance department survive let alone turn it into a profit center? One of the tools is the use of a CMMS. A properly selected CMMS can tremendously help your organization.

Following is a list of benefits that can be derived using a CMMS. You have to review your organization and see what applies.

Increased labor productivity. If the system provides the employees with a planned job, the procedures, needed parts, and tools, the employees will be able to go directly to the job and do the needed work with no delays or interruptions. The employee will also work safer, since job plans would include all safety procedures. All this will increase labor productivity.

Efficient asset management and maintenance scheduling will contribute to reduced overtime. Improved parts tracking and availability will reduce the unproductive time of employees.

Overall you will witness:

1. Reduction of overtime.
2. Reduction of outside contract work.
3. Reduced maintenance backlog.
4. Reduced cost per repair.
5. Improved morale of employees by diffusing employee frustration; a happy worker is a productive worker.
6. Better service to other departments.
7. A significant reduction in paperwork to make the most productive use of employee time.
8. Effective utilization of maintenance and supervisory personnel's time.
9. Reduced follow-up role required of the supervisor.

Increased equipment availability. Within a few months of implementation, it will become much easier to identify repetitive faults and trends. This information will assist in maximizing equipment uptime and reducing breakdowns. Your emphasis should shift from reactive to proactive maintenance, and your ratio of percent planned to unplanned jobs should increase. A corresponding reduction in downtime will follow.

Savings through reduced production loss results primarily from performing adequate *predictive maintenance* (PdM) and PM. PdM employs sensors or detectors to monitor equipment performance and condition. Detection of potential problems prompts the writing of the needed work order. For example, when an unacceptable reading of bearing temperature or vibration is sensed, a maintenance response can automatically initiate. The response may take the form of an automatic injection of lubricant or an initiation of a maintenance work order.

This PdM helps keep the equipment in good condition. It leads to timely repairs rather than waiting for an actual failure to initiate corrective maintenance action.

PM is the regular scheduling of specific maintenance tasks to prevent possible anticipated failures. PM tasks include work such as a filter or bearing replacement, and calibration and condition checking.

An effective PM program will keep equipment in good condition because it forces periodic monitoring, and it serves as an early detection system for finding problems before they mature into full failures. The immediate result is that maintenance jobs are kept to minimum size. The long-term result is equipment that retains its effectiveness, value, and reliability in supporting production.

Longer useful life of equipment. You can prolong the effective lifetime of your assets and equipment through regular, adequate PM. The CMMS will support the processes involved in prolonging the life of your assets. It also improves resale value of the equipment. Overall you will witness reduction of downtime.

Inventory control

Reduced inventory costs. Planning of jobs permits parts to be available when and where needed. Experience shows that a reduction of 10 to 15 percent in parts stocked and consumed is possible. Reductions also extend to inventory carrying charges, as well as to stockroom size, staff, and service requirements.

As work becomes more predictive, so does your stock holding. Carrying out regular stock reviews allows you to minimize stock and to reduce expensive inventory. Spare parts can be linked to equipment, ensuring that obsolete parts are readily identified. Many users find that the greatest returns from a CMMS come through improvements in inventory control, with savings of 10 to 15 percent being typical.

Product quality. Improved product quality results primarily from performing adequate PdM and PM. PdM and PM details are described in the "Equipment availability" section earlier. Good equipment condition assures good product quality.

Environment control

Safety issues. Preventing accidents and injuries as a result of proper procedures documented by CMMS can save you a significant amount of money.

Compliance issues. Compliance with industry regulations. Some industries such as food processing, pharmaceuticals, and petrochemicals require that your asset management systems comply with the national or international standards that regulate their industry. If you require such compliance then you should select a CMMS that has a provision for the same. Meeting the regulatory requirements can save you money that you would otherwise pay in fines for not meeting the requirements.

Maintenance information. Access to maintenance information is dramatically expanded. Other benefits are also expected of a CMMS. Improved reporting and support for management control can contribute strongly to justifying a CMMS. Crucial production line decisions can be simplified by dependable and timely data on equipment condition, and expected lifetime. Such information can provide guidance in setting the size of production runs, deciding on equipment replacement, and pricing the product.

Better management of service contracts. Service contracts are arrangements with outside contractors for continuing services such as fork truck maintenance. Because these services are managed maintenance, it is convenient to use the CMMS to manage the functionality and accounting part of the services.

Blanket purchase orders. Blanket purchase orders with parts and materials suppliers may be managed effectively by a CMMS. This is an important area for tight control as blanket purchase orders are proven to be major leakages in the maintenance budget.

Increase in overall plant productivity. Review of all these benefits indicates that a CMMS can help increase overall plant/facility productivity. Figure 8.1 shows key elements of CMMS benefits.

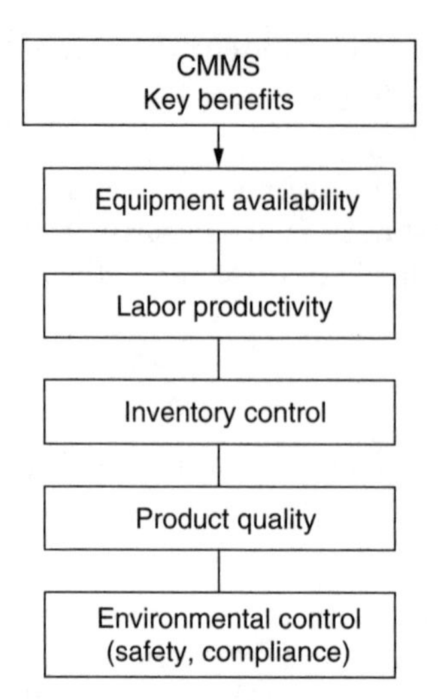

Figure 8.1 CMMS benefits (key elements).

Profit-Driven Maintenance (PDM)

Profit-driven maintenance (PDM) is an approach to bridge the gap between executive and maintenance management with the primary goal of maximizing organizational profits. You just have to learn to speak the language management speaks.

For example: Presenting a project to the management where you are going to reduce breakdowns, downtime, and the like may not be well received. The same project will be very well received if it is projected to increase sales capacity. The ultimate goal still remains the same for every one: increasing overall productivity and profits for the company.

Conclusion

Most organizations view maintenance as an expense and a cost center. A properly implemented CMMS can help maintenance cope with downsizing and turn maintenance into a profit center. It is important to bridge the communication gap between executive and maintenance management to increase overall productivity.

Bibliography

Bagadia, K., *Microcomputer-Aided Maintenance Management*, Marcel Dekker, Inc., New York, 1987.

Bagadia, K., "Talking Shop with Software Suppliers," *PIMA*, November 1988, pp 45–46.

Bagadia, K., "Selecting CIMM," *Manufacturing Sys.*, December 1985, pp 24–32.

Bagadia, K., "Deciding What You Want Your System to Do," *PIMA*, September 1988.

Bagadia, K., "Justifying the Cost to Your Boss,"

Burress, K., "How Reliability Affects Earnings Per Share,"

"CMMS Implementation Survey Results-2004," at www.plant-maintenance.com.

CMMS User Handbook, Thomas Publishing Company, New York, 1993.

Dave, B., "Maximum CMMS," at www.plant-maintenance.com.

Danforth, H., "Communication and Support Through CMMS," *Facility Manage.*,

Davis, D., and J. Mikes, "Reaping the Benefits of CMMS," *Maintenance World*, August 2003.

Dunn, S., "Implementing a Computerized Maintenance Management System," at www.plant-maintenance.com.

Evans, R., "Too small for a CMMS? Think Again," *Maintenance Technol.*,

Graham, P., "The Other Road Ahead," at www.paulgraham.com.

Hamilton, S., "Workforce Automation: Mobile Computing for Asset Management," at www.CMMSCity.com.

Hartmann(Ed.), "Prescription for Total TPM Success," *Maintenance Technol.*,

Hemming, R. J., and D. L. Davis, "Eat an Elephant-Implement a CMMS," *Maintenance World*, July 2003.

Idhammer, "CMMS and Preventive Maintenance (Part 1)," at www.idcon.com.

Idhammer, "CMMS and Preventive Maintenance (Part 2),"at www.idcon.com.

Idhammer, "CMMS and Preventive Maintenance (Part 3)," www.idcon.com.

Kennedy, S., "Managing Spare Parts," *Maintenance World*, March 2004.

Klein, B., "Preparing to Implement CMMS at Your Facility," *Facility Manage.*,

Koslow, N., "Ownership, Accountability Give Employees Power," *Maintenance Technol.*

Leonard, J. L., "Internships: An Under-Utilized Maintenance Resource," *Maintenance World.*

Levitt, J., "50 Questions to Help Your CMMS Search," at www.plant-maintenance.com.

MacMillan, S., and M. Lance, "Managing An EAM/CMMS Project: Phase 2," *Maintenance Technol.*,

MacMillan, S., and M. Lance, "Managing An EAM/CMMS Project: Phase 3," *Maintenance Technol.*,

Mason, E. R., "Justifying Maintenance Projects," *Maintenance Technol.*, August 1989.

Mather, D., "Backlog Management," *Maintenance World*

Mather, D., "Training-The Backbone of Cultural Change," at www.maintenanceworld.com.

Mather, D., "Developing CMMS Implementation Templates," *Maintenance World*, September 2003.

Mitchell, J., "Profit or Cost Center Mentality: Is the Difference Important," *Maintenance Technol.*

Oas, A., "Did you control the chaos?," *Maintenance World*, September 2003.

Oberg, P., "Managing Maintenance as a Business," at www.maintenanceworld.com.

Ralph, P., "The CMMS/EAM Benchmarking System," The Maintenance Excellence Institute,

Thomas, C., "Maintenance-Business Center Approach,"

Vesier, C., "A Six-Step Work Process to Increase Profitability with Reliability Improvements,"
Vujicic, A., "How to Avoid Becoming Another CMMS Implementation Failure Statistic," at www.CMMSCity.com.
Weir, B., "Failure Codes," *Maintenance World*
Westerkamp, T., "Auditing for CMMS Success," at www.facilitiesnet.com.
Williamson, R., "TPM: An Often Misunderstood Equipment Improvement Strategy," *Maintenance Technol.*,
Williamson, R., "The Basic Pillars of Total Productive Maintenance," *Maintenance Technol.*.
Winston, C., "Administration and Training: Keys to CMMS Implementation Success," at www.plant-maintenance.com.

Glossary

Words and Terms

AIDC Automatic identification and data capture.

ADC Automatic data capture.

AIT Automatic identification technology.

Application program Software having a specific usage (e.g., CMMS and CAD).

Bar code A means for designating (coding) alphabetic characters or numbers by a series of lines of varying width (bars) to be read by an optical scanner.

CMMS Computerized maintenance management system—a set of integrated software that performs functions in support of the management of maintenance operations.

CST Client server technology.

Condition-based maintenance (CBM) Maintenance based on actual conditions in a facility based on in-house measurements.

Computer integrated facility management (CIFM) Also known as *computer aided facility management* (CAFM). A computer system that manages usable space, furniture, fixtures, and equipment in a facility.

Condition monitoring (CM) Recording various building characteristics such as temperature, electrical usage, ventilation, vibration, and other parameters.

Corrective/improvement work orders Work orders describing work and resources needed for one time correction of a problem, or for making a specific improvement.

Data entry validation Automatic checking of specific characteristics of entered data to ensure it is within predefined standards.

Data record A single set of the data in a database (e.g., a single work order within the work order database).

Downtime The period during which production is halted due to equipment unavailability.

Format A specific on-screen or printed arrangement of data (e.g., a work order form).

Function Any activity performed by software (e.g., in a CMMS, the checking of parts availability, or rescheduling of preventive maintenance work).

GIS Geographic information system.

GPS Global positioning system.

Hardware The computers, printers, and other devices on which software programs run.

IT Information technology.

LAN Local area network.

MIS Management information system.

MRO Maintenance, repair, operations.

Maintenance procedures library A set of descriptions of the standard job steps and resources needed for performing specific maintenance jobs.

Menu driven User choices are made from on-screen lists of options.

Module Discrete segments of software performing one or more specific functions; may be added or removed at user discretion with most CMMSs (e.g., preventive maintenance, work order).

O/M Operations and maintenance.

PDA Personal digital assistance.

PLC Programmable logic controller.

Password security Prevention from entering and using a software system unless a valid password is entered. This may include selective restriction of the use of the system's various operating sections.

Pick list A list of stock room shelf items to be collected for use in performing a maintenance work order.

Planning Describing the job steps and resources such as labor, parts, and support equipment, required for a work order.

Preventive maintenance Predefined sets of maintenance tasks performed to avoid equipment failures and breakdowns.

Predictive maintenance (PdM) Maintenance work called for by the results of (sensor) measurements of specific equipment operating characteristics, which indicate possible failures.

RF Radio frequency.

RFID Radio frequency identification device.

Reliability-centered maintenance (RCM) A methodology used to identify probable system and equipment failures and to increase plant safety.

Report generator Computer software (preferably integral part of CMMS), which allows creation of a report from any set of data selected from a database.

Scheduling Fixing the date and time for performing a maintenance job, having assured the availability of the resources described in the planning process.

Sensor A device whose measurement of heat, vibration, electrical, or other physical characteristics will be used as input data for predictive maintenance.

Service contract A contract with an outside supplier of a regularly provided maintenance service such as fork-lift truck maintenance.

Software The programs and databases that run on computer hardware.

Source code Programs in their original programming language.

WAN Wide area network.

CMMS Module Definitions

Bar-code reading and printing Ability to use bar codes in parts, work order, and equipment identification. This capability permits work orders, equipment numbers, stock numbers, and other identifying numbers to be translated into bar codes. Use of those codes increases the accuracy in matching all the mentioned identifiers so that the correct items are used on the correct job and on the correct piece of equipment.

Cost and budgeting Development and management of cost estimates and budgets for maintenance projects and their distribution to various other plant departments. Data managed in this module would include the details of the maintenance budget and current cost experience. Reports would be designed and used for control of maintenance department expenses.

Equipment data Listing and basic data for equipment owned and maintained. Information recorded may include equipment number, location, maintenance priority, cost, manufacturer, model and serial numbers, and the dates of purchase and installation. Equipment parts lists and supporting information such as calibration data, may also be kept in this module.

Inventory control Recordkeeping of maintenance parts and other materials received, stocked, and disbursed as well as their locations in the stockroom. This module provides the information needed for managing the stock of parts and materials used in performing maintenance work. Data is provided showing what items are to be ordered, what are available, and what are on order. The data so provided is directly supportive of the planning and scheduling effort.

Labor Recordkeeping of employee information. Data on employees would provide the maintenance scheduler with the roster of available employees and their skills and training for job assignment purposes.

Maintenance procedure library A compilation of standard procedures to be used on work orders for specific jobs. Job step activities and labor, parts and materials required are listed. This serves as a guide to the work order planner and is updated whenever a standard procedure is changed.

Planning and scheduling Manages the detailed planning of the labor, materials, and all other resources needed to complete a work order. Permits describing the specific job steps, required labor types, and estimated times as well as all other resources such as tools and equipment for performing corrective and improvement work orders. Planned work orders for which all resources are available may then be scheduled.

Predictive maintenance Equipment monitoring devices are used to determine current conditions such as vibration frequency and amplitude or bearing temperature. Readings are interpreted by software/hardware devices to determine whether or not a problem exists and what is its nature. If a problem is found, a corrective work order is written for the needed work. Any or all parts of this process may be automated.

Preventive maintenance Manages work orders for predefined repetitive scheduled jobs in order to preserve good equipment condition. Generates work orders and automatically schedules and records completions. Jobs done under PM usually include inspections, lubrications, and changes of finite-lifetime items such as filters or seals. An important assignment of PM is discovering needed corrective work, which results in generating the needed work orders.

Purchasing Creation and processing of purchase orders. This module manages the purchasing function beginning with the automatic creation of a purchase order when the reorder point of a stock item is reached. Data usually recorded includes a list of vendors and prior purchase information. Reports available would include current items on order and vendor status and history.

Work orders Manages work orders for correcting faults or improving equipment's condition. Issues, edits, and records work orders as history. Basic work order information on the desired job might include the equipment number to be worked on, a description of the problem or work to be done, who requested the work, when they wanted it completed, the job's priority, and the date issued. Once the basic information is entered, the system records the work order and supports editing the details on work needed, material required, and other pertinent data. Planning and scheduling of work orders may be available. The completion of the work order may be recorded and the work order placed in a history file. Reporting may be available on work orders in current backlog as well as in the history file.

Index

ABOUT THE AUTHOR

Kishan (Kris) Bagadia, a leading expert on computerized maintenance management systems, provides CMMS consulting to manufacturing and healthcare organizations and conducts maintenance seminars around the world. He is a frequent speaker at industrial organizations such as the Society of Manufacturing Engineers, Association of Facilities Engineering, and Manufacturing Week Show and Conference, as well as at a number of universities. He makes complex topics easy to understand. He can be contacted at krisb@peakis.com.